Rafał A. Bachorz

Implementation and Application of the Explicitly Correlated Coupled-Cluster Method in Turbomole

Implementation and Application of the Explicitly Correlated Coupled-Cluster Method in Turbomole

by
Rafał A. Bachorz

universitätsverlag karlsruhe

Dissertation, Universität Karlsruhe (TH)
Fakultät für Chemie und Biowissenschaften
Tag der mündlichen Prüfung: 06.07.2009
Referenten: Prof. Dr. W. Klopper, PD Dr. Karin Fink

Impressum

Universitätsverlag Karlsruhe
c/o Universitätsbibliothek
Straße am Forum 2
D-76131 Karlsruhe
www.uvka.de

Universitätsverlag Karlsruhe 2009
Print on Demand

ISBN: 978-3-86644-392-1

Implementation and application of the explicitly correlated coupled-cluster method in TURBOMOLE

Zur Erlangung des akademischen Grades eines

Doktors der Naturwissenschaften

(Dr. rer. nat.)

der Fakultät für Chemie und Biowissenschaften
der Universität Karlsruhe (TH)
angenommene

Dissertation

von
Magister

Rafał A. Bachorz

aus Poznań, Polen

Dekan:	Prof. Dr. S. Bräse
1. Gutachter:	Prof. Dr. W. Klopper
2. Gutachter:	PD Dr. Karin Fink
Tag der mündlichen Prüfung:	6. Juli 2009

Contents

1 Introduction

Quantum chemistry, an important part of chemistry, over the last years has become a branch of science that is capable of predicting, confirming and sometimes rejecting experimental observations. A very good example of the latter might be an achievement of the Polish scientists, Kołos and Wolniewicz. In 1960, using an approach based on an explicitly correlated wave function, they predicted a dissociation energy of hydrogen molecule that was in disagreement with the measurement of Herzberg [1], who was an unquestioned authority in the field at that time. The subsequent experiment, carried out by Herzberg again, gave an excellent agreement with the theoretical prediction. This was a victory of quantum chemistry and, in general, theoretical sciences.
The main goal of quantum chemistry is to find a solution of the molecular Schrödinger equation [2]

$$\hat{H}\Psi = E\Psi, \tag{1}$$

where $\hat{H}$ is the electronic molecular Hamiltonian, Ψ is the wave function and E is the energy. Due to the fact the we are unable to solve this equation exactly, we have to apply various approximations. In principle the molecular wave function Ψ can be constructed from N-electron Slater determinants and thus, when treating Eq. (1) in an approximated manner, we face two major problems. One of them is the incompleteness of the one-electron basis used to represent the molecular orbitals (MOs), the other problem refers to the approximations made within the N-electron model. The errors that come from these two sources are of different nature, but to large extend depend on each other. Schematically they are illustrated on Figure 1. The perfect quantum-chemistry model would give

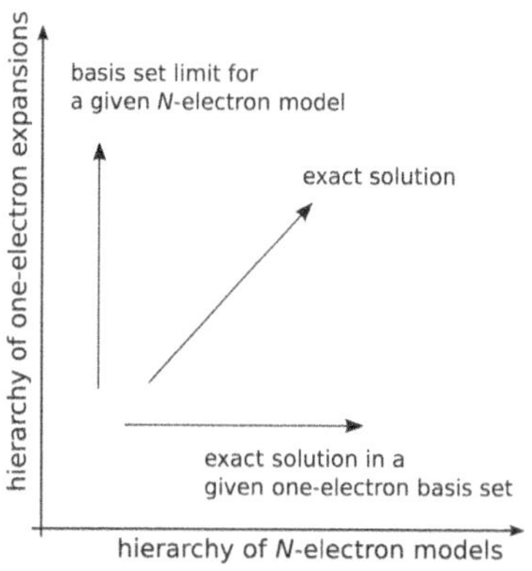

Figure 1: The systematic approach to the exact solution of the Schrödinger equation.

us the solution that is located in the upper-right corner of the plot (denoted here as "exact solution"). Within the computational practice, however, we develop approximated models that approach the exact solution. The final tool, which is the practical realization of the assumed model, gives us the results that are almost always a compromise between accuracy and computational cost.
The incompleteness of the one-electron basis set can be reduced in two ways. One can apply either extrapolation techniques or, alternatively, one can include into the wave function terms that depend explicitly on the distance between the electrons. The latter solution seems to be more satisfactory since the enhanced, explicitly correlated wave

function gives a much improved description of the system. Extrapolation, on the other hand, improves only the energy based on the analytical model for the convergence of that quantity.

The improvements within the hierarchy of N-electron models is probably a more complicated task. There are a couple of quantum-chemistry theories that allow us to approach the exact Schrödinger equation systematically. Among them the coupled-cluster (CC) method represents probably the most successful approach. It can be applied to relatively large systems and the theory is both size-extensive and size-consistent. So far the only way to approach the exact Schrödinger equation, within the hierarchy of the N-electron models (following the horizontal axis on Figure 1), is the systematic extension of the excitation level. In CC theory there is a series of models that refer to the way the cluster operator is truncated (CCS, CCSD, CCSDT, CCSDTQ and so on). In the limit of the untruncated cluster operator the CC wave function becomes equivalent with full CI, which is the exact solution of the Schrödinger equation within a particular basis set. The truncation level indicates, in some sense, the accuracy of the model which is almost always limited by the available computational resources.

The main goal of the present work is to report the implementation of the explicitly-correlated coupled-cluster singles-and-doubles method (CCSD(F12)) in TURBOMOLE. This tool is capable of very efficient calculating the CCSD energies at the basis set limit with relatively small orbital basis sets. The implementation works with RHF, UHF and ROHF reference wave functions, which means that it can treat both closed- and open-shell species. The formulation in terms of intermediate quantities and the application of density fitting techniques make this implementation quite unique.

The thesis begins with Section 2, where a brief history about the explicitly correlated approaches is presented. This is followed by Section 3 with general remarks about standard and explicitly correlated coupled-cluster theories. In Section 4, the details about the CCSD(F12) model relevant to the implementation in TURBOMOLE are presented. The usefulness of the developed tool is illustrated with the application to the problems that are of interest to general chemistry. A very accurate determination of the reactions barrier heights of two $CH_3^\bullet + CH_4$ reactions has been carried out (Section 5) and the atomization energies of 106 medium-size and small molecules were computed and compared with available experimental thermochemical data (Section 6). The ionization potentials and electron affinities of the atoms H, C, N, O and F were obtained and an agreement with the experimental values of the order of a fraction of a meV was reached (Section 7). Within all applications, the CCSD(F12) calculation was only a part of the whole computational procedure. The contributions from various levels of theory were taken into account to provide the final result, that could be successfully compared to the experiment.

2 Explicit correlation

The slow convergence of configuration interaction (CI) expansions in orbital basis sets is linked to the presence of the correlation cusp in the wave function. Within the molecular Hamiltonian the interelectronic Coulomb operator scales like the reciprocal of the distance between the electrons and, for the part of the configuration space where the electrons are close to each other, the Coulomb interaction diverges. However, the local energy defined as

$$E_{\mathrm{loc}} = \frac{\hat{H}\Psi}{\Psi} \tag{2}$$

remains constant everywhere for Ψ being the exact wave function. The singularity that comes from the Coulomb operators must be cancelled by the kinetic energy operator. This can happen only if the wave function becomes linear in r_{12} as the particles coalesce. The careful examination of this problem by Kato [3] led to the well known cusp condition

$$\left(\frac{\partial\Psi(r_{12})}{\partial r_{12}}\right)_{r_{12}\to 0,\mathrm{av}} = \frac{1}{2}\Psi(r_{12}=0), \tag{3}$$

where "av" indicates a spherical averaging. The cusp condition in a more general form was given by Pack and Brown [4]. The incorrect behavior of the wave functions based on the conventional CI expansion for $r_{12} = 0$ is the major reason why these expansions converge slowly with respect to the orbital expansion. In fact, when considering the energies, the small sphere of the configuration space around the coalescence point contributes very little to this quantity. More important for energies is an overall shape and size of the Coulomb hole. This problem is illustrated on Figure 2, where wave functions of the electronic ground state of helium atom are plotted. In the upper part (a), the difference between the Hylleraas and Hartree-Fock wave functions is shown, where one of the electrons is fixed in the plane that contains the nucleus, at a position 0.5 a_0 away from the nucleus. The Hylleraas wave function can be considered here as the exact solution to the Schrödinger equation for helium. In the middle part (b) the geminal function is plotted, in this case the summation over xy simplifies to just one term because for helium two electrons occupy one orbital . The bottom part (c) represents the difference between (a) and (b). This difference is very small, which means that the F12 geminal functions describe properly the correlation hole.

2.1 Hylleraas and James-Coolidge wave functions

Already in 1929 Hylleraas found that the orbital expansion of the wave function of helium converges extremely slowly. The problem could be overcome by including terms in the wave function that depend explicitly on interelectronic coordinates [6, 7]. The proposed explicitly correlated wave function was of the form

$$\Psi = \exp(-\alpha s)\sum_{klm} c_{klm} s^k t^{2l} u^m \tag{4}$$

with the following definition of the coordinates

$$s = r_1 + r_2, \tag{5}$$

$$t = r_1 - r_2, \tag{6}$$

$$u = r_{12}. \tag{7}$$

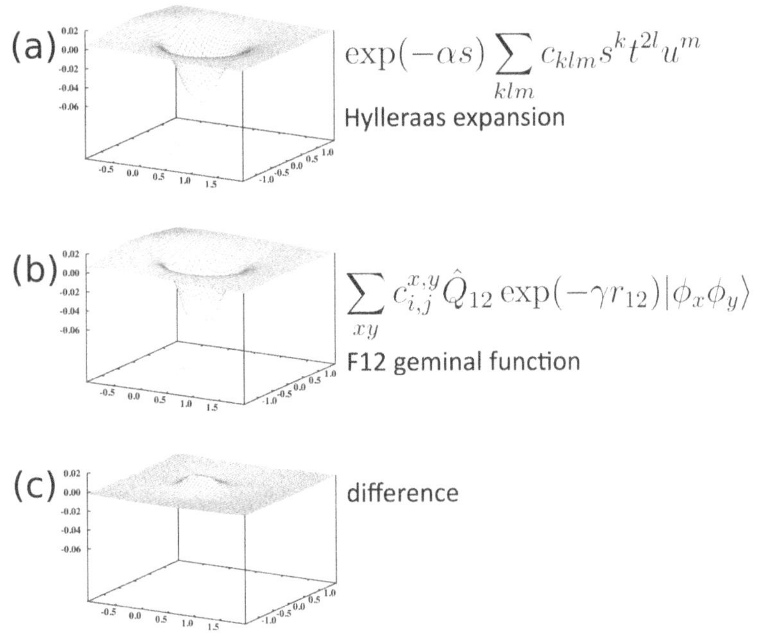

Figure 2: The wave function of helium atom in its electronic ground state. The upper part (a) represents the difference between the Hartree-Fock and Hylleraas wave functions in a plane that contains the nucleus and fixed electron, called in the literature the Coulomb hole [5]. In the middle part (b) the F12 geminal function with $\gamma = 1.0\ a_0^{-1}$ is plotted. The bottom part (c) represents the difference between (a) and (b).

The singlet ground state (para-helium case) requires particular spatial symmetry of the function. Therefore, only even powers of t were considered here. Even very short expansions of this type [Eq. (4)] led to very good results reducing the discrepancy between theory and experiment from 0.12 to 0.01 eV in terms of the ionization potential of the helium atom.
A natural idea was to extend the concept of Hylleraas to molecules. It was done in 1933 by James and Coolidge [8], and the generalized function was of the form

$$\Psi = \sum_{i=1}^{n} c_i \left[\phi_i(\boldsymbol{r}_1, \boldsymbol{r}_2) + \phi_i(\boldsymbol{r}_2, \boldsymbol{r}_1)\right], \tag{8}$$

with the following definition of the basis functions for the Σ state

$$\phi_i(\boldsymbol{r}_1, \boldsymbol{r}_2) = \exp(-\alpha(\xi_1 + \xi_2))\xi_1^{m_i}\xi_2^{n_i}\eta_1^{k_i}\eta_2^{l_i}\rho^{\mu_i} \tag{9}$$

were ξ and η are the elliptic coordinates

$$\xi_i = \frac{r_{ai} + r_{bi}}{R}, \quad \eta_i = \frac{r_{ai} - r_{bi}}{R}, \tag{10}$$

r_{ai} is the distance between the nucleus a and the i-th electron, R is the internuclear distance, ρ is the explicitly correlated term $\rho = 2r_{12}/R$, α is the nonlinear parameter to be optimized and m_i, n_i, k_i, l_i, μ_i are non-negative integers. Similar to the Hylleraas case, even very short expansions of the type Eq. (8) led to very accurate results yielding the total energy of H_2 $E = -1.17347\ E_h$, which is ca. 1 mE_h above the variational limit. This result was later improved by Kołos and Roothaan [9], and Kołos and Wolniewicz [10, 11]. The wave function Eq. (8) does not have the proper behavior in the dissociation limit and the idea of Kołos and Wolniewicz was to correct this drawback. A new, generalized wave function was introduced as

$$\Psi = \sum_{i=1}^{n} c_i \left[\phi_i(\boldsymbol{r}_1, \boldsymbol{r}_2) + \phi_i(\boldsymbol{r}_2, \boldsymbol{r}_1)\right], \tag{11}$$

with the basis functions

$$\phi_i(\boldsymbol{r}_1, \boldsymbol{r}_2) = \exp(-\alpha\xi_1 - \bar{\alpha}\xi_2)\cosh(\beta\eta_1 + \bar{\beta}\eta_2)\xi_1^{m_i}\xi_2^{n_i}\eta_1^{k_i}\eta_2^{l_i}\rho^{\mu_i}. \tag{12}$$

The definition of the coordinates and used symbols is analogous to Eq. (9). Using this function, called Kołos-Wolniewicz wave function, and correcting the electronic energy with adiabatic and relativistic contributions, Kołos and Wolniewicz obtained a dissociation energy of the hydrogen molecule that was in disagreement with the contemporary experimental value [1]. Subsequent experiment, however, removed the discrepancy [12], proving the validity of the quantum mechanical approach. The achievement of Kołos and Wolniewicz is considered as a great triumph of quantum chemistry.

2.2 Exponentially correlated Gaussians (ECG)

The necessary condition for the effectiveness of the quantum mechanical approach is the possibility of expressing the integrals in an analytical way. For small linear systems it was possible in the case of the Kołos-Wolniewicz wave function [Eq. (11)], but the generalization to non-linear many-electron cases led to a catastrophic complication of the integrals. In 1960, Boys [13] and Singer [14] independently proved that for a particular class of functions, that were the products of Cartesian Gauß functions and Gauß correlation factors, all the integrals can be computed in an analytical way. The general wave function expressed in terms of the corresponding ECG basis functions can be written as

$$\Psi = \hat{\mathcal{A}}\left(\Theta_{S,M_s}\hat{\mathcal{S}}\sum_i c_i\phi_i\right), \tag{13}$$

where $\hat{\mathcal{A}}$ and $\hat{\mathcal{S}}$ are permutational and spatial symmetry operators, respectively. The Θ_{S,M_s} is a proper spin eigenfunction and the basis functions are defined as

$$\phi_i = \left[\prod_{k=1}^{N} g\left(\alpha_{k_i}, \boldsymbol{A}_{k_i}, l_{k_i}, m_{k_i}, n_{k_i}\right)\right]\left[\prod_{\gamma=1}^{N-1}\prod_{\delta=\gamma+1}^{N} e^{-\beta^i_{\gamma\delta}r^2_{\gamma\delta}}\right], \tag{14}$$

$$\begin{aligned} g\left(\alpha_{k_i}, \boldsymbol{A}_{k_i}, l_{k_i}, m_{k_i}, n_{k_i}\right) = {} & \left(x_k - A^x_{k_i}\right)^{l_{k_i}}\left(y_k - A^y_{k_i}\right)^{m_{k_i}}\left(z_k - A^z_{k_i}\right)^{n_{k_i}} \\ & \times \exp(-\alpha_{k_i}|\boldsymbol{r}_k - \boldsymbol{A}_{k_i}|^2), \end{aligned} \tag{15}$$

where N is number of electrons in the considered system. The centers of the Gaussian functions $\boldsymbol{A}_{k_i}$ as well as the nonlinear parameters α_k, $\beta_{\gamma\delta}$ are variationally optimized and l_k, m_k, n_k are non-negative integer numbers. The integrals can be expressed through the exponential function, square root, a finite number of the basic algebraic operations and the well known in the literature Boys function

$$F_m(t) = \int_0^1 \exp(-tx^2)x^{2m}dx. \tag{16}$$

The special case of ECG functions, referring to two-electron systems, is called Gaussian-type geminal (GTG). The mathematical form of the GTG basis function can be written as

$$\phi_i = \prod_{k=1,2} g\left(\alpha_{k_i}, \boldsymbol{A}_{k_i}, l_{k_i}, m_{k_i}, n_{k_i}\right) \exp(-\beta_i r_{12}^2). \tag{17}$$

Expansions based on GTG basis functions [Eq. 17] have been used as Ansätze for the explicitly correlated MP2 and CC wave functions [15].
It is worth to mention that each basis function within the ECG expansion [Eq. (14)] correlates all electrons, which yields in total $n_{\text{bas}}(3N + N(N+1)/2)$ nonlinear parameters to be optimized, where n_{bas} is the number of the basis functions in the expansion. Each ECG basis function depends on the coordinates of all electrons, but the resulting $3N$-dimensional integrals can be computed in closed form. The accuracy obtained with ECGs is unprecedented [15], but due to the time-consuming nonlinear optimization the high accuracy could be obtained only for few-electron atoms and molecules.

2.3 The R12 methods

The common feature of the explicitly correlated approaches discussed so far is that the whole wave function is expanded in explicitly correlated basis functions. Kutzelnigg and Klopper proposed a different approach, initially at the MP2 level of theory [16]. The general idea of Kuztelnigg and Klopper was to supplement the conventional CI expansion with the explicitly correlated part in the following way

$$|\text{R12}\rangle = |\text{CI}\rangle + \tfrac{1}{2}\hat{\mathcal{A}} \sum_{k<l} r_{kl} |\text{Hartree}\rangle, \tag{18}$$

where $\hat{\mathcal{A}}$ is the antisymmetry operator, r_{kl} is the distance between the electrons k and l and $|\text{Hartree}\rangle$ is the Hartree state (*i.e.* the product of occupied orbitals). The prefactor "$\frac{1}{2}$" in Eq. (18) has been formally introduced to fulfill the Kato's electron-electron cusp condition [3] [see Eq. (3)]. The intrinsic feature of the MP2 approach is the possibility of decomposing the correlation energy into pair contributions, provided that the pair functions are strongly orthogonal on the occupied space [17, 18]. Therefore, if a reference wave function of the form Eq. (18) is assumed, the entire MP2 pair function can be written as

$$|u_{ij}\rangle = |w_{ij}\rangle + |v_{ij}\rangle. \tag{19}$$

The pair function Eq. (19) contains the conventional part expanded in products of the virtual orbitals

$$|w_{ij}\rangle = \sum_{ab} t_{i,j}^{a,b} |ab\rangle, \tag{20}$$

and the explicitly correlated part written as

$$|v_{ij}\rangle = \sum_{xy} c_{i,j}^{x,y} \hat{Q}_{12} f_{12} |xy\rangle. \tag{21}$$

These equations are expressed in the spin-orbital formalism and the products of orbitals are assumed to be antisymmetrized. The coefficients $c_{i,j}^{x,y}$ are the explicitly correlated analogues of the conventional $t_{i,j}^{a,b}$ amplitudes. The xy indices refer to the space of geminal replacements which is usually spanned by the occupied orbitals. The operator $\hat{Q}_{12}$ in Eq. (21) is the strong orthogonality projector and f_{12} is the correlation factor. In Eq. (18) the f_{12} correlation factor was chosen as linear r_{12} term. It is not necessary to use it in such form. Recent advances in R12 theory have shown that Slater-type correlation factors, referred here as f_{12}, are advantageous. Depending on the choice of the Ansatz of the wave function, the formula for the projector varies, but the detailed discussion of these issues is postponed until Subsection 4.2. The minimization of the Hylleraas functional

$$H[u_{ij}] = \langle u_{ij}|(\hat{f}_1 + \hat{f}_2 - \epsilon_i - \epsilon_j)|u_{ij}\rangle + 2\langle u_{ij}|r_{12}^{-1}|ij\rangle \geq \epsilon_{ij}^{(2)} \tag{22}$$

with respect to the amplitudes yields upper bounds to the MP2 pair energies. The presence of the operator $\hat{Q}_{12}$ in the explicitly correlated pair function [Eq. (21)] leads to three- and four-electron integrals, *e.g.* within the second term of the Hylleraas functional [Eq. (22)]

$$\langle ij|r_{12}^{-1}\hat{O}_1 f_{12}|xy\rangle = \sum_m \langle ijm|r_{12}^{-1} f_{23}|mxy\rangle, \tag{23}$$

where $\hat{O}_1$ is projection operator onto the space spanned by the occupied orbitals. The six-index integrals [Eq. (23)], even though not very difficult to evaluate, are numerous, but due to the insertion of the resolution-of-the-identity (RI) operators they can be avoided. The advised formulation of the approximated RIs (using, *e.g.* the complementary auxiliary basis set (CABS) approximation) keeps the associated errors small. For instance they could be reduced to the order of 0.01 mE_h in the correlation energy for systems of the size of Ne [19]. The details about the approximated RIs, relevant to the present work, are discussed later in Section 4.2.

Modern implementations of the MP2-F12 method combine the CABS approximation [20] with *robust* density fitting techniques [21, 22] and local approaches [23]. The $c_{i,j}^{x,y}$ coefficients are usually constrained at the values predetermined from the cusp conditions, as one half for singlet pairs and one quarter for triplet pairs in the spin-adapted formalism [24, 25]. The MP2-F12 methods have been extended to treat open-shell systems with unrestricted [26, 27, 28], restricted [29, 30] and multireference [29] formalisms.

The inclusion of explicit correlation at the MP2 level of theory can be relatively easily done by supplementing the conventional pair function [Eq. (20)] with the explicitly correlated part [Eq. (21)], such that the entire pair function is the sum of these two [Eq. (19)]. At the coupled-cluster level of theory an analogous modification of the wave function is achieved by modifying the cluster operator with an additional term that depends explicitly on the interelectronic distance. The resulting coupled-cluster residuals and energy expression, supplemented with the explicitly correlated contributions, lead to significant enhancements of the energy and the wave function. It is possible to obtain quintuple-ζ quality energies with triple-ζ orbital basis sets [31, 32]. The main focus of the present work is to report the theory and the practical aspects of the implementation of the CCSD(F12) method.

3 Coupled-cluster theory

Within the standard quantum-chemistry approach we begin our description with solving the Hartree-Fock (HF) equations. This is a mean-field approach and the solution, *i.e.* an electronic state, is represented by the set of occupied molecular orbitals. The expectation value of the molecular Hamiltonian with the Hartree-Fock wave function is an approximation to the ground state energy of the system [33, 34]

$$E_{\mathrm{HF}} = \sum_i \langle i|\hat{h}|i\rangle + \frac{1}{2}\sum_{ij}\langle ij||ij\rangle, \tag{24}$$

where $\langle ij||ij\rangle$ is an antisymmetrized Coulomb integral, and $\hat{h}$ is a one-electron part of the molecular Hamiltonian. The Hartree-Fock energy [Eq. (24)] does depend only on the occupied orbitals (denoted with $i, j, k, \ldots$). The virtual orbitals $(a, b, c, \ldots)$, that are the eigenvectors of the Fock operator, might be considered as the "byproducts" of the HF procedure. In contrast to the occupied states they are not directly optimized, but they are obtained via the diagonalization of the Fock matrix. Therefore the subspace of the virtual orbitals is qualitatively different from the occupied orbitals. This remark is relevant here, because all post HF methods involve the excited Slater determinants where at least one occupied MO is replaced with a virtual MO.
The HF picture does not account for dynamic electron correlation. In fact, within formal definition of the correlation energy

$$E_{\mathrm{corr}} = E_{\mathrm{exact}} - E_{\mathrm{HF}}, \tag{25}$$

the HF approach is assumed as entirely insensitive for electron correlation. Therefore the HF description has to be improved. In standard quantum chemistry the refinement is obtained by introducing virtual electronic excitations into a set of virtual MOs obtained also in the HF procedure (*vide supra*). In the limit of arbitrary high virtual excitations and complete virtual orbital spaces, the exact solution of the Schrödinger equation can be obtained. It means that the exact solution of the Schrödinger equation requires the full configuration interactions treatment carried out in an infinite basis set. This is of course impossible and, instead, we must deal with an approximated description of the electronic system. In particular we must skip some multiple electronic excitations by truncating the excitation level, and the space of virtual orbitals must be finite. The approximations made within these two components of the quantum chemistry model determine its accuracy and performance. These fundamental sources of errors have been already mentioned in the introductory section (see Section 1).

3.1 Standard coupled-cluster theory

The Hartree-Fock wave function is obtained by variational minimization of the energy functional with respect to orthonormal transformations of the orbitals

$$\langle \mathrm{HF}|\hat{H}|\mathrm{HF}\rangle = \min_{\Psi_{\mathrm{det}}}\langle \Psi_{\mathrm{det}}|\hat{H}|\Psi_{\mathrm{det}}\rangle. \tag{26}$$

The energy functional involves the Slater determinant, that is an antisymmetrized product of occupied spin orbitals. As the solution, the optimal wave function and the associated energy are obtained.

In a given orbital basis, the Hartree-Fock description divides the orbital space into a set of occupied and virtual spin orbitals. From the Slater determinant any other determinant may be generated by replacing an occupied orbital by a virtual. Formally such operation is performed within the second quantization formalism by using the excitation operators

$$\hat{\tau}_i^a = a_a^\dagger a_i, \quad \hat{\tau}_{ij}^{ab} = a_a^\dagger a_i a_b^\dagger a_j, \quad \hat{\tau}_{ijk}^{abc} = a_a^\dagger a_i a_b^\dagger a_j a_c^\dagger a_k, \quad \ldots . \tag{27}$$

An operator a_i annihilates an electron in spin orbital i, while $a_a^\dagger$ creates an electron in spin orbital a. The creation and annihilation operators satisfy the usual algebra $[a_p, a_q^\dagger]_+ = \delta_{pq}$. Introducing the generic notation $\hat{\tau}_\mu$ for excitation operators and noting that $[\hat{\tau}_\mu, \hat{\tau}_\nu] = 0$, we may express the coupled-cluster wave function in the standard exponential form [35]

$$|\mathrm{CC}\rangle = \exp(\hat{T})|\mathrm{HF}\rangle, \tag{28}$$

where the cluster operator

$$\hat{T} = \sum_\mu t_\mu \hat{\tau}_\mu, \tag{29}$$

is a linear combination of excitation operators $\hat{\tau}_\mu$, each multiplied by an associated cluster amplitude t_μ. To determine these amplitudes and calculate the electronic energy, we insert the wave function Eq. (28) into the Schrödinger equation

$$\hat{H}|\mathrm{CC}\rangle = E|\mathrm{CC}\rangle \tag{30}$$

and multiply from the left with $\exp(-\hat{T})$, yielding the similarity-transformed Schrödinger equation

$$\exp(-\hat{T})\hat{H}\exp(\hat{T})|\mathrm{HF}\rangle = E|\mathrm{HF}\rangle. \tag{31}$$

To determine the cluster amplitudes and the electronic energy we multiply this equation from the left by the ground state (Hartree-Fock state $\langle \mathrm{HF}|$) and all possible excited states, called projection manifolds

$$\langle \mu| = \langle \mathrm{HF}|\hat{\tau}_\mu^\dagger. \tag{32}$$

The projections against all possible states that might be generated by the excitation operators present in the cluster operator Eq. (29) according to Eq. (32) give the recipe how to compute the coupled-cluster state from the following nonlinear equations

$$E = \langle \mathrm{HF}|\exp(-\hat{T})\hat{H}\exp(\hat{T})|\mathrm{HF}\rangle, \tag{33}$$

$$0 = \langle \mu|\exp(-\hat{T})\hat{H}\exp(\hat{T})|\mathrm{HF}\rangle. \tag{34}$$

The coupled-cluster electronic state is uniquely defined by the set of the cluster amplitudes t_μ, and these amplitudes are used to obtain the coupled-cluster energy from Eq. (33). Due to the fact that the Ansatz of the coupled-cluster wave function has the exponential parametrization [Eq. (28)] the energy is size-extensive. This is an obvious advantage of the coupled-cluster formalism compared to some other techniques (*e.g.* configuration interaction). For a general discussion of coupled-cluster theory and the coupled-cluster equations see Refs. [5, 36].
With all possible excitations included in the cluster operator Eq. (29), coupled-cluster theory provides an exact solution to the Schrödinger equation in a given orbital basis,

equivalent to a full configuration-interaction (FCI) treatment in the same basis. However, except in a very small orbital basis, this approach is very expensive and the approximations have to be applied. The approximate treatments can be straightforwardly obtained by truncating the cluster operator Eq. (29). Introducing the notation

$$\hat{T} = \hat{T}_1 + \hat{T}_2 + \hat{T}_3 + \cdots, \tag{35}$$

where $\hat{T}_n$ contains all possible n-excitation operators, the following hierarchy of the coupled-cluster models can be introduced

$$|\text{CCS}\rangle = \exp(\hat{T}_1)|\text{HF}\rangle, \tag{36}$$

$$|\text{CCSD}\rangle = \exp(\hat{T}_1 + \hat{T}_2)|\text{HF}\rangle, \tag{37}$$

$$|\text{CCSDT}\rangle = \exp(\hat{T}_1 + \hat{T}_2 + \hat{T}_3)|\text{HF}\rangle, \tag{38}$$

and so on. Thus, at the lowest level of coupled-cluster singles [CCS, Eq. (36)] theory, we include all possible single excitations in the cluster operator; in coupled-cluster singles-and-doubles [CCSD, Eq. (37)] theory [37], all possible double excitations are also included. In coupled-cluster singles-doubles-triples [CCSDT, Eq. (38)] theory [38, 39], all triple excitations are included. In general, higher-order excitations are less important than the lower-order ones, except that the singles play a relatively minor role, effectively performing small orbital adjustments to the doubles and higher excitations.
It is important to note that, at each level of coupled-cluster theory, we include through the exponential parameterization of Eq. (28) all possible determinants that can be generated within a given orbital basis, that is, all determinants that enter the FCI wave function in the same orbital basis. Thus, the improvement in the sequence CCSD, CCSDT, and so on does not occur because more determinants are included in the description but from an improved representation of their expansion coefficients. For example, in CCS theory, the doubly-excited determinants are represented by $\frac{1}{2}\hat{T}_1^2|\text{HF}\rangle$, whereas the same determinants are represented by $(\hat{T}_2 + \frac{1}{2}\hat{T}_1^2)|\text{HF}\rangle$ in CCSD theory. Thus, in CCSD theory, the weight of each doubly-excited determinant is obtained as the sum of a connected doubles contribution from $\hat{T}_2$ and a disconnected singles contribution from $\hat{T}_1^2/2$. This parameterization of the wave function is not only more compact than the linear parameterization of configuration-interaction (CI) theory, but it also ensures size-extensivity of the calculated electronic energy.

3.2 Explicitly correlated CCSD theory

The hierarchy of coupled-cluster models provides a clear route towards the exact solution of the Schrödinger equation, but the slow basis-set convergence limits the accuracy sometimes even for small molecules. The way to overcome this problem is to combine the coupled-cluster model with the explicitly correlated approach. It can be done, in principle, for any model within the coupled-cluster hierarchy. The main task of this work is, however, the implementation of explicitly correlated CCSD model, hence the discussion will be focused on this particular model.
The main idea of explicitly correlated CCSD theory is to extend the conventional space of excitations such that pairs of occupied orbitals are replaced by an explicitly correlated geminal function

$$g_{xy}(1,2) = \sum_{\alpha\beta} w_{xy}^{\alpha\beta} \phi_\alpha(1)\phi_\beta(2). \tag{39}$$

where a formally complete, one-electron basis $\{\alpha, \beta \ldots\}$ has been introduced. The additional excitations are included through the generalization of the cluster operator [Eq. (29)]. The conventional double excitations $(i, j \rightarrow a, b)$ are extended with the additional ones that refer to the complete space $(i, j \rightarrow \alpha, \beta)$. The operator replacing the pair of occupied orbitals by the explicitly correlated geminal functions can be written as

$$\hat{T}_{2'} = \frac{1}{2} \sum_{xiyj} c_{i,j}^{x,y} \sum_{\alpha\beta} w_{x,y}^{\alpha,\beta} \hat{E}_{\alpha i} \hat{E}_{\beta j}, \tag{40}$$

where $\hat{E}_{\alpha i}$ are excitation operators in the spin-free notation (see Subsection 4.2). The geminal function Eq. (39) can be considered as the analogue of the MP2-F12 pair function Eq. (21). The explicitly correlated coupled-cluster equations are derived by inserting $T_{2'}$ into the usual expressions for the coupled-cluster residuals and energy. The amplitudes $c_{i,j}^{x,y}$ are determined in a manner analogous to conventional amplitudes, that is, the similarity-transformed Schrödinger equation is projected onto the manifold of geminal double replacements

$$\sum_{\alpha\beta} w_{x,y}^{\alpha,\beta} \left\langle \overline{\begin{matrix} \alpha\beta \\ ij \end{matrix}} \middle| \exp(-\hat{T}) \hat{H} \exp(\hat{T}) \middle| \mathrm{HF} \right\rangle = 0. \tag{41}$$

Similar to the MP2-F12 formalism, the strong orthogonality projector in the geminal basis leads to many-electron integrals in the amplitude equations. Explicit evaluation of these integrals severely restricts the range of application and the successful approaches are those that involve two-electron integration at most. The implementation of the CCSD(F12) in TURBOMOLE fulfills this requirement.

4 The CCSD(F12) model in TURBOMOLE

4.1 General remarks

One of the special cases of coupled-cluster theory is the singles-and-doubles (CCSD) model [37]. The cluster operator Eq. (29) is restricted to contain only the singles and doubles excitation operators. The importance of this model can be seen from the fact that, for any coupled-cluster wave function, the singles and doubles amplitudes are the only ones that contribute directly to the coupled-cluster energy. In the explicitly correlated CCSD model the conventional cluster operator containing the $\hat{T}_1$ and $\hat{T}_2$ operators is supplemented with an additional term that takes care of the explicit correlation (written with red font)

$$|\text{CCSD-F12}\rangle = \exp(\hat{T}_1 + \hat{T}_2 + \hat{T}_{2'})|\text{HF}\rangle = \exp(\hat{S})|\text{HF}\rangle, \tag{42}$$

where $|\text{HF}\rangle$ is the Hartree-Fock state and the $\hat{T}_1$, $\hat{T}_2$ and $\hat{T}_{2'}$ are the excitation operators. The CCSD-F12 equations can be obtained by projecting the function

$$\exp(-\hat{S})\hat{H}\exp(\hat{S})|\text{HF}\rangle \tag{43}$$

onto the Hartree-Fock state and the excited determinants [see Eqs. (33) and (34)]. The new term in the cluster operator requires an additional projection onto the manifold of geminal double replacements [see Eq. (41)]. This yields the following equations for the CCSD-F12 model [40]

$$E = \langle\text{HF}|\exp(-\hat{S})\hat{H}\exp(\hat{S})|\text{HF}\rangle, \tag{44}$$

$$0 = \langle\mu_1|\exp(-\hat{S})\hat{H}\exp(\hat{S})|\text{HF}\rangle, \tag{45}$$

$$0 = \langle\mu_2|\exp(-\hat{S})\hat{H}\exp(\hat{S})|\text{HF}\rangle, \tag{46}$$

$$0 = \langle\mu_{2'}|\exp(-\hat{S})\hat{H}\exp(\hat{S})|\text{HF}\rangle. \tag{47}$$

The derivation of the working expression can be done with the use of Wick's theorem, a diagrammatic approach, or any other technique [5, 36]. The expressions of the CCSD(F12) model, an approximation to the full CCSD-F12 method [41], can be obtained by neglecting some terms in the formulas of the CCSD-F12 model. In particular, the terms quadratic in the $c_{i,j}^{x,y}$ coefficients and higher order in the fluctuation potential vanish within the CCSD(F12) approach [41]. The CCSD(F12) equations, expressed in terms of the similarity-transformed Hamiltonian $\bar{H} = \exp(-\hat{T}_1 - \hat{T}_2)\hat{H}\exp(\hat{T}_1 + \hat{T}_2)$, can be written in the following way

$$E = \langle\text{HF}|\bar{H} + [\bar{H}, \hat{T}_{2'}]|\text{HF}\rangle, \tag{48}$$

$$0 = \langle\mu_1|\bar{H} + [\bar{H}, \hat{T}_{2'}]|\text{HF}\rangle, \tag{49}$$

$$0 = \langle\mu_2|\bar{H} + [\bar{H}, \hat{T}_{2'}]|\text{HF}\rangle, \tag{50}$$

$$0 = \langle\mu_{2'}|\bar{H} + [\hat{f}, \hat{T}_{2'}]|\text{HF}\rangle, \tag{51}$$

where $\hat{f}$ is the Fock operator. The main purpose of this Section is to provide detailed information about working expressions, derived from Eqs. (48)-(51), which have been implemented in TURBOMOLE as main objective of this thesis.

The current Section is organized as follows: in Subsection 4.2 the general information about the excitation operators, strong orthogonality projectors and correlation factors is provided. In Subsection 4.3 the general sketch of the program flow is shown, where the most important steps of the implementation are shortly discussed. In Subsection 4.4 the details about the $V^{\mu,\nu}_{x,y}$ intermediate are provided. This quantity is one of the most important intermediates within the F12 part of the CCSD code, and a significant fraction of the terms that contribute to the residuals is obtained from this quantity. Therefore, the discussion about $V^{\mu,\nu}_{x,y}$ precedes the subsections that focus on explicitly correlated contributions to the coupled-cluster energy (Subsection 4.5) and residuals (Subsections 4.6, 4.7, 4.8). In the last part of this Section, the alternative approximations to the CCSD-F12 model, also implemented in TURBOMOLE as part of the thesis, are discussed (Subsection 4.9).

4.2 Excitation operators, strong orthogonality projectors and correlation factors

Before analyzing the final formulas of the CCSD(F12) model, it is beneficial to discuss some technical details related to the excitation operators. The implementation in TURBOMOLE supports restricted, unrestricted and restricted open-shell Hartree-Fock reference wave functions (RHF, UHF and ROHF, respectively). The calculations based on the latter reference are treated in the UHF fashion with the use of semicanonical orbitals [42, 43]. Depending on the choice of the reference, the excitation operators (as well as the other quantities) are expressed in the spin-free or spin-orbital formalisms, respectively for the RHF and UHF cases. In the spin-orbital formalism the spin indices are included explicitly and therefore one has to consider various spin cases separately. For instance, in the case of two-index quantities there are two spin cases (α and β), four-index quantities require considering four spin cases ($\alpha\alpha$, $\beta\beta$, $\alpha\beta$ and $\beta\alpha$). However, due to the fact that we can interchange the coordinates of the electrons without affecting any expectation value, the $\beta\alpha$ and $\alpha\beta$ cases become equivalent and only one of them has to be considered explicitly. There are special cases where such symmetry cannot be applied, *e.g.* the spin off-diagonal components ($\alpha\beta$ and $\beta\alpha$) of the $f^{p'',q}_{x,y}$ integrals, where the upper indices belong to different spaces of orbitals. Quantities that have more than four indices pose similar properties, but they do not occur in the CCSD(F12) model. In the RHF limit, that is when the spatial parts of the α and β orbitals are the same, one can sum over spin cases leaving only the orbital indices. The single and double excitations operators in this case, called excitation operators in the spin-free notation, can be written in the following way

$$\hat{T}_1 = \sum_{ia} t^a_i \hat{E}_{ai}, \tag{52}$$

$$\hat{T}_2 = \frac{1}{2} \sum_{aibj} t^{a,b}_{i,j} \hat{E}_{ai} \hat{E}_{bj}, \tag{53}$$

with the definition of the spin-free $\hat{E}_{ai}$ operator

$$\hat{E}_{ai} = \sum_{\sigma} a^{\dagger}_{a\sigma} a_{i\sigma}. \tag{54}$$

Detailed information about excitation operators and their properties can be found in [5]. Within the spin-orbital formalism, which is appropriate for the formulas based on the UHF reference, both spin and orbital indices are included explicitly. The analogous excitation operators in this case have the following structure

$$\hat{T}_1 = \sum_{\sigma} \sum_{ai} t_{i\sigma}^{a\sigma} \times \hat{\tau}_{i\sigma}^{a\sigma}, \tag{55}$$

$$\hat{T}_2 = \frac{1}{2} \sum_{\sigma\sigma'} \sum_{aibj} t_{i\sigma,j\sigma'}^{a\sigma,b\sigma'} \times \hat{\tau}_{i\sigma,j\sigma'}^{a\sigma,b\sigma'}. \tag{56}$$

It is worth to mention that in the UHF formalism, due to the orthogonality of the spin functions, it is not allowed to excite an electron occupying α spin-orbital to the β spin-orbital (and vice versa). As it was already mentioned in Subsection 3.2, in the case of explicitly correlated CCSD theory, the cluster operator is supplemented with the additional excitation operator $T_{2'}$ [Eq. (40)]. This operator is responsible for the explicitly correlated treatment and involves additional excitations (F12 excitations) into the complementary basis. In the spin-free formalism its mathematical form was already shown and briefly discussed [Eq. (40)], in the spin-orbital formalism this operator can be introduced as

$$\hat{T}_{2'} = \frac{1}{2} \sum_{\sigma\sigma'} \sum_{xiyj} c_{i\sigma,j\sigma'}^{x\sigma,y\sigma'} \sum_{\alpha\beta} w_{x\sigma,y\sigma'}^{\alpha\sigma,\beta\sigma'} \hat{\tau}_{i\sigma,j\sigma'}^{\alpha\sigma,\beta\sigma'}. \tag{57}$$

where the indices $\alpha\beta$ **do not** refer here to the spin but to the complete virtual basis. The space of the conventional virtual orbitals is extended with additional functions

$$\{\phi\} = \{\phi_i\} \oplus \{\phi_\alpha\} = \{\phi_i\} \oplus \{\phi_a\} \oplus \{\phi_{\alpha_\perp}\}. \tag{58}$$

The space $\{\phi_{\alpha_\perp}\}$ is an algebraic complement of the space spanned by the occupied and virtual orbitals to a full (complete) space $\{\phi\}$ (see Figure 3). Practically one can use

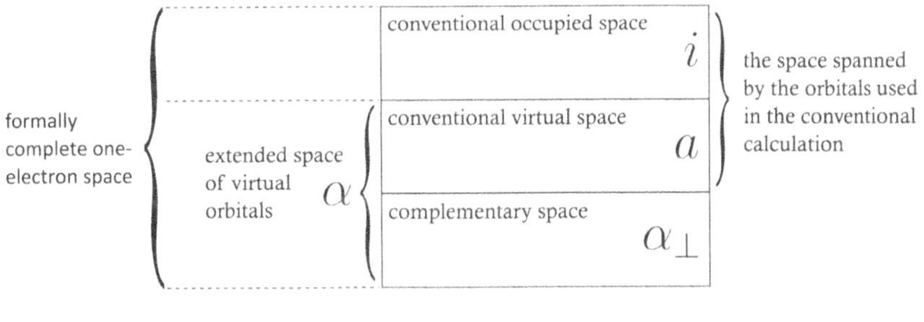

Figure 3: The inclusion of the complementary space into the space of conventional MOs.

the complementary auxiliary basis set approximation (CABS) proposed by Valeev [20], which means that the space of conventional MOs is supplemented with additional CABS orbitals. These additional orbitals, obtained as linear combination of both the conventional and the auxiliary basis functions, are chosen in such a way that CABS orbitals are orthonormal with respect to the conventional MOs and also mutually orthonormal (the CABS approximation will be discussed in more detail later in this Section). The coefficients for the representation of the geminal functions $w_{x,y}^{\alpha,\beta}$ are the matrix elements (ignoring spins)

$$w_{x,y}^{\alpha,\beta} = \langle \alpha\beta | \hat{Q}_{12} f_{12} | xy \rangle. \tag{59}$$

The presence of the projector $\hat{Q}_{12}$ assures that the space spanned by the conventional double excitations is strongly orthogonal on the occupied space. Within the current work two Ansätze for $\hat{Q}_{12}$ are considered. To be consistent with the literature we will refer to "Ansatz 1" and "Ansatz 3" (abbreviated as **An1** and **An3**, respectively). The f_{12} term is an arbitrary function of the distance between the electrons r_{12}. Within the current work the linear r_{12} and Slater type correlation factors $\exp(-\gamma r_{12})$ are considered. Due to the computational convenience, the latter one was implemented as a linear combination of Gaussian functions (LCG) in the following way

$$\exp(-\gamma r_{12}) = \sum_{i=1}^{N} c_i \exp(-\alpha_i r_{12}^2), \tag{60}$$

where the γ parameter is kept constant and the linear c_i, and non-linear α_i coefficients are determined in a least-squares manner [44]. The CCSD(F12) electronic energies treated with the orbital basis set of triple-ζ (or higher) quality do not depend strongly on the value of γ [31, 44]. In practice we normally choose some value between 0.9 and 1.4 a_0^{-1}. The length of the expansion, N, is also not critical, usually six terms assure appropriate accuracy. It is worth to note here that it is not necessary to use the approximated Slater correlation factor. Ten-no derived the analytical formulas with the exact Slater function, but they are much more demanding from the implementation point of view [25, 45]. As it was mentioned above, the presence of the $\hat{Q}_{12}$ operator is the intrinsic feature of the explicitly correlated geminal function and its choice determines the particular Ansatz of the wave function. In the case of **An1**, the projector has the following form [46]

$$\hat{Q}_{12}^1 = (\hat{1} - \hat{P}_1)(\hat{1} - \hat{P}_2), \tag{61}$$

where $\hat{P}_1$ and $\hat{P}_2$ are the one-electron projectors onto the space of orthonormal spatial orbitals that are contained in the finite basis set (spanned by conventional occupied and virtual orbitals)

$$\begin{aligned} \hat{P}_1 &= \sum_p |\phi_p\rangle\langle\phi_p|, \\ \hat{P}_2 &= \sum_q |\phi_q\rangle\langle\phi_q|. \end{aligned} \tag{62}$$

The indices p and q in Eq. (62) indicate that the orbitals ϕ_p and ϕ_q refer to electrons 1 and 2, respectively. The standalone operators $\hat{P}_1$ and $\hat{P}_2$ within the operator $\hat{Q}_{12}^1$

$$\hat{Q}_{12}^1 = \hat{1} - \hat{P}_1 - \hat{P}_2 + \hat{P}_1\hat{P}_2, \tag{63}$$

give rise to three electron integrals. This can be easily shown in the case of the V intermediate, the quantity that is very often used to express the explicitly correlated terms (see Subsection 4.4 for details). The difficult term is the following

$$\langle pq|f_{12}\hat{P}_1 r_{12}^{-1}|st\rangle = \langle pqr|f_{12}r_{23}^{-1}|rts\rangle, \tag{64}$$

whereas the analogous term that contains the product of the operators $\hat{P}_1\hat{P}_2$ factorizes

$$\langle pq|f_{12}\hat{P}_1\hat{P}_2 r_{12}^{-1}|tu\rangle = \langle pq|f_{12}|rs\rangle\langle rs|r_{12}^{-1}|tu\rangle, \tag{65}$$

where the indices p, q, r, s, t, u refer to arbitrary orbitals. The many-electron integrals [Eq. (64)] would be avoided if the standalone operators were not present in the formula. This can be achieved via a procedure called standard approximation (SA), which essentially is the multiple insertion of the resolution-of-the-identity (RI) into the projector

$$\hat{Q}_{12}^{1} = \hat{1} - \hat{P}_1\hat{1}_2 - \hat{1}_1\hat{P}_2 + \hat{P}_1\hat{P}_2. \tag{66}$$

In the limit of the complete basis the representation of the unity operator is exact

$$\hat{1}_1 = \sum_{p=1}^{\infty} |\phi_p\rangle\langle\phi_p|, \tag{67}$$

but in practice we have to use an approximated resolution-of-the-identity expanded in a particular basis. This leads to further approximations and ambiguities. If the orbital basis set is chosen to represent the $\hat{1}$ operator

$$\hat{1}_1 \approx \hat{P}_1, \quad \hat{1}_2 \approx \hat{P}_2 \tag{68}$$

then the entire $\hat{Q}_{12}^{1}$ projector simplifies to

$$\hat{Q}_{12}^{1} = \hat{1} - \hat{P}_1\hat{1}_2 - \hat{1}_1\hat{P}_2 + \hat{P}_1\hat{P}_2 \approx \hat{1} - \hat{P}_1\hat{P}_2 - \hat{P}_1\hat{P}_2 + \hat{P}_1\hat{P}_2 = \hat{1} - \hat{P}_1\hat{P}_2. \tag{69}$$

The advantage of this approach is its simplicity, but one has to use relatively large orbital basis sets to ensure that the approximated unity operators in Eq. (68) are well represented [22]. This limitation quickly becomes a bottleneck and apparently such approach gives reliable results only for relatively small systems (treated with large orbital basis sets). A natural idea is to introduce an additional basis set (different than the orbital one) that is used to represent the resolution-of-the-identity operators. There are two (so far) closely related models, the auxiliary basis set approximation (ABS) [47] and the complementary auxiliary basis set approximation (CABS) [20]. In the implementation in TURBOMOLE only the latter one was considered. Within this approach the whole basis (singly primed), used for the representation of the unity operators, is the union of the orbital and some auxiliary basis (doubly primed)

$$\{p'\} = \{p\} \cup \{p''\}. \tag{70}$$

The main idea of the CABS approach is that the RI operators are represented by using the extended basis set [Eq. (70)]

$$\hat{P}_1' = \sum_{p'} |\phi_{p'}\rangle\langle\phi_{p'}| = \hat{P}_1 + \hat{P}_1'' \approx \hat{1}_1. \tag{71}$$

The expression for the $\hat{1}_2$ operator can be obtained in complete analogy. The projector [Eq. 66)] can now be rewritten as

$$\begin{aligned}
\hat{Q}_{12}^{1,\mathrm{CABS}} &= \hat{1} - \hat{P}_1\hat{1}_2 - \hat{1}_1\hat{P}_2 + \hat{P}_1\hat{P}_2 \approx \hat{1} - \hat{P}_1\hat{P}_2' - \hat{P}_1'\hat{P}_2 + \hat{P}_1\hat{P}_2 && (72)\\
&= \hat{1} - \hat{P}_1(\hat{P}_2 + \hat{P}_2'') - (\hat{P}_1 + \hat{P}_1'')\hat{P}_2 + \hat{P}_1\hat{P}_2 && (73)\\
&= \hat{1} - \hat{P}_1\hat{P}_2 - \hat{P}_1\hat{P}_2'' - \hat{P}_1\hat{P}_2 - \hat{P}_1''\hat{P}_2 + \hat{P}_1\hat{P}_2 && (74)\\
&= \hat{Q}_{12}^{1} - \hat{P}_1\hat{P}_2'' - \hat{P}_1''\hat{P}_2. && (75)
\end{aligned}$$

The quantities obtained with **An1** involve the general and CABS indices. It is important to note that we avoid here the quantities having two CABS indices simultaneously. The **An1** has been implemented in the TURBOMOLE program, but the results yielding by this Ansatz are nowadays considered as less accurate than those obtained with **An3**.
The most successful Ansatz in F12 theory uses the following form of the strong orthogonality projector [19, 20]

$$\hat{Q}_{12}^{3} = (\hat{1} - \hat{O}_1)(\hat{1} - \hat{O}_2) - \hat{V}_1\hat{V}_2. \tag{76}$$

This choice refers to **An3**. $\hat{O}_1$ and $\hat{V}_1$ are the one-electron projectors onto the orthonormal spatial orbitals that are included in the occupied and virtual spaces, respectively

$$\hat{O}_1 = \sum_i |\phi_i\rangle\langle\phi_i|, \tag{77}$$

$$\hat{V}_1 = \sum_a |\phi_a\rangle\langle\phi_a|. \tag{78}$$

All formulas with the projector Eq. (76) have been derived and implemented in TURBOMOLE in the course of this thesis. To avoid many electron integrals one can, similar to the case of **An1**, apply the SA procedure with the use of the CABS approximation. This yields the following contributions

$$\begin{aligned} \hat{Q}_{12}^{3} \approx& \hat{1} - \hat{O}_1\hat{P}_2 - \hat{O}_1\hat{P}_2'' + \hat{P}_1\hat{O}_2 + \hat{P}_1''\hat{O}_2 + \hat{O}_1\hat{O}_2 - \hat{V}_1\hat{V}_2 \\ =& \hat{1} - \hat{P}_1\hat{P}_2 - \hat{O}_1\hat{P}_2'' - \hat{P}_1''\hat{O}_2 = \hat{Q}_{12}^{1} - \hat{O}_1\hat{P}_2'' - \hat{P}_1''\hat{O}_2. \end{aligned} \tag{79}$$

Graphically, the action of this projector is explained in Figure 4. The space spanned by

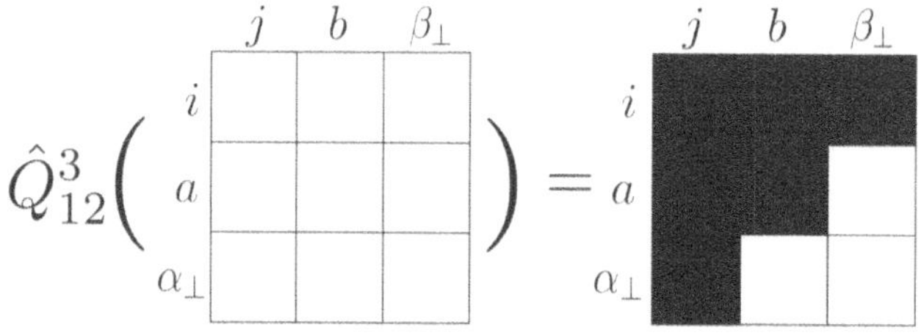

Figure 4: The pictorial representation of the projector within Ansatz 3.

the conventional excitation $(i, j \to a, b)$ is separated out and only the geminal excitations involving the CABS orbitals are left. Hence, after the action of operator Eq. (79), only the double replacements of the form $(i, j \to \alpha_\perp, \beta_\perp)$, $(i, j \to a, \beta_\perp)$ and $(i, j \to \alpha_\perp, b)$ remain. The other excitations, referring to the conventional replacements, are not considered within the explicitly correlated calculation. The indices $\alpha_\perp$ and $\beta_\perp$ refer to the complementary space that is orthogonal to the space of the conventional MOs.

4.3 The general sketch of the program flow

The implementation of the CCSD(F12) model in TURBOMOLE involves many quantities and a variety of algorithms that are used for the evaluation of the CCSD(F12) energy.

The present work should be considered as the extension and generalization of the existing implementation of the conventional CCSD model [48]. The manner, in which the CCSD, and thus CCSD(F12), models are realized can be characterized as integral-direct density-fitting implementation. It means that some quantities are computed in the density-fitting approximation (DF), the other ones are computed on-the-fly in the AO basis and contracted with some other quantities [48]. The DF approximation is in the literature often called the RI approximation (resolution-of-the-identity), but within the current work the term "RI approximation" is reserved for the intrinsic RI used in the R12 theory. The main idea of the DF approximation relies on the assumption that the four-index quantities are never stored explicitly. Instead, they are computed on-the-fly from three-index integrals by carrying out the contraction over the index associated with the auxiliary basis set. The index of the DF auxiliary basis (usually denoted with Q) should be clearly distinguished from the CABS index (denoted as doubly primed index p''), they refer to entirely different spaces. The DF approximation requires additional well optimized basis sets, but, especially at the MP2 level of theory, the speed up of the program compensates the effort related to this extra basis set. Within the series of the many-body models, MP2 as the simplest one that takes care of dynamic correlation, is a bit unusual. Within this model the bottleneck is the computation of the integrals, the contraction between the integrals and the amplitudes is a much cheaper step. This is the reason for the significant speed up of the code if the DF approximation is applied for the four-index Coulomb integrals at the MP2 level of theory. The scaling of the higher many body models (CCSD, CCSDT and so on) is determined by the cost associated with the contractions. For these models (especially for those that go beyond the CCSD model) the DF approximation is not that critical.

The implementation of the CCSD(F12) model in TURBOMOLE relies on the built-in MP2-F12 module [49, 50]. The general sketch of the algorithm is shown in Algorithm 1.

Result: The CCSD(F12) energy

1 Perform the MP2-F12 calculation (involved step, described in detail elsewhere)
2 Compute the integrals over the $(f_{12}r_{12}^{-1})$ operator with two AO indices
3 Transform the MP2-F12 quantities (call to **ccr12_prep** from the main program **ricc2**)
4 Resort the integrals (call to **rir12_resort** from **ccsdint** routine)
5 Carry out the back-transformation of the four-index integrals (call to **ccr12_pqtomunu** from **ccsdint** routine)
6 Compute the non T_1-dependent V intermediate (call to **rir12ccsd_vint** from **ccsdint** routine, described in detail in Subsection. 4.4)
7 Transform the $V_{x,y}^{\mu,\nu}$ intermediate to the occupied space $V_{x,y}^{i,j}$ (call to **cc_bfmo** from **ccsdint** routine)
8 Prepare the three-index quantities $B_{Q,ap''}$ (**ccsdint** routine)
9 Compute the Fock matrix elements $F_{ip''}$ and $F_{ap''}$ (**ccsdint** routine)
while *Convergency criteria not fulfilled (energy, norm of the T_1 residual)* **do**
10 Compute the T_1, T_2 and $T_{2'}$ residuals (described in detail in Subsections. 4.7, 4.8)
11 Compute the energy (**cc_energy**)
12 Precondition the amplitudes (**cc_precond**)
13 Get the DIIS guess for the next-iteration amplitudes (**cc_diis**)
end

Algorithm 1: General algorithm of the CCSD(F12) module in the TURBOMOLE program.

Each CCSD(F12) calculation is preceded with the MP2-F12 call (step 1; Alg. 1). The whole MP2-F12 energy calculation is performed and the important intermediates are saved on disk for later use within the CCSD(F12) calculation. The MP2-F12 step is quite involved by itself but will not be discussed here. The details can be found in Refs. [49, 50]. From the point of view of the present work, the important information is that the MP2-F12 intermediates are later on used within the CCSD(F12) module. They are collected in Table 1. For practical reasons the integrals $(fg)_{x,y}^{\mu,\nu}$, although not needed for the evaluation of the MP2-F12 energy, are computed within the MP2-F12 module (step 2; Alg. 1). These quantities are used for the computation of the V intermediate if the LCG correlation factor is requested. After the MP2-F12 run the first CCSD(F12)

Table 1: The MP2-F12 quantities used in the CCSD(F12) calculation.

Description	quantity	purpose
MP2-F12 matrices :	$B_{ij}^{kl}, X_{ij}^{kl}, Q_{ij}^{kl}$	the $T_{2'}$ residual, preconditioning
four-index integrals :	$f_{ij}^{pq}, f_{ij}^{p''q}$	the T_2, $T_{2'}$ residuals
three-index integrals :	$B_{Q,kl}, B_{Q,kp''}$	all residuals, Fock matrices
two-index quantities :	$C_{\mu,p''}, C_{\mu'',p''}$	back-transformation

routine (**cc_prep**) is responsible for preliminary issues (step 3; Alg. 1). One of them is transforming the spin adapted singlet/triplet MP2-F12 matrices (Table 1) into full (ij, ji) matrices. Also the f_{12} integrals (Table 1) are transformed to the form that is appropriate for later use in the CCSD(F12) part. The next part is the resorting of the four-index quantities to the form that is suitable for the CCSD(F12) part. The different storing schemes within the MP2-F12 and CCSD(F12) codes refers to the fact that the implementation of the many-body models in the **ricc2** program is designed for the cases where a relatively large system is treated with a rather small, or medium-size basis set. This is reflected in the structure of the code, in particular in the storing schemes. All four-index quantities (integrals, amplitudes) are stored in such a way, that the occupied indices run the fastest. This is not the case for the MP2-F12 quantities, therefore the resorting procedures have to be called at the preliminary level of the CCSD(F12) run. Let us introduce the following convention that describes the storing scheme of the quantities

$$A(\text{the fastest} \ldots \text{the slowest}) \qquad e.g.: A(p, q, r, s). \tag{80}$$

In the above example the quantity A is stored in such a way that the index p is the fastest, the next one is q and so on. Using this convention the resorting shown in step 4 of Algorithm 1 can be written as

$$f_{12}(p, q, i, j) \longrightarrow f_{12}(i, j, p, q), \tag{81}$$

where the quantity $f_{12}(p, q, i, j)$ is computed and stored at the MP2-F12 level. The resorting of the other integrals is always similar, the occupied indices that run the slowest at the MP2-F12 level are exchanged with the general (or CABS) ones. From the point of view of the implementation such procedure is not entirely trivial. A naive routine would assume that the whole four-index quantity is kept in the memory. For larger systems (or larger basis sets) this assumption would become very quickly a bottleneck, therefore an algorithm that batches the integrals into the pieces that fit into the available memory was designed. The next step is the back-transformation of the four-index integrals which is needed for the calculation of the $V_{x,y}^{\mu,\nu}$ intermediate. This part of the code (step 5; Alg. 1) is explained in Subsection 4.4 where the issues associated with this intermediate are discussed. When the $V_{x,y}^{\mu,\nu}$ is computed (step 6; Alg. 1) the next task is to transform this quantity to the occupied space (step 7; Alg. 1). The $V_{x,y}^{i,j}$ is needed for the evaluation of the CCSD(F12) energy (which will be explained in Subsection 4.5) and the transformation can be written as

$$V_{x,y}^{i,j} = \sum_{\mu\nu} (C_{\mu i})^{\dagger} V_{x,y}^{\mu,\nu} C_{\nu j}. \tag{82}$$

It was already mentioned that within the current implementation the geminal indices (xy) refer to the occupied space. At step 8 of Algorithm 1 the additional three-index quantities are computed. The preliminary run of the MP2-F12 program provides the integrals shown in Table 1, the additional integrals, needed at the CCSD(F12) level, have the orbital indices in the CABS and virtual spaces, $B_{Q,ap''}$. They are needed for the computation of the valence component of the Fock matrix elements (*vide infra*) and the C and D contributions to the T_2 and $T_{2'}$ vector functions (see Subsection 4.8 for details). The detailed information about the implementation of the three-index integrals in TURBOMOLE can be found in [51]. The last step, before starting the CC iterations, refers to the computation of the valence parts of the Fock matrices involving the CABS index (step 9; Alg. 1). The adjective "valence" here means that only the active two-electron part of the Fock operator is included. These Fock matrices are computed within the DF approximation and can be written in the following way

$$F^{val}_{ip''} = \sum_{k} \{2^{\mathrm{DF}} g^{p'',k}_{i,k} - {}^{\mathrm{DF}} g^{k,p''}_{i,k} \}, \tag{83}$$

$$F^{val}_{ap''} = \sum_{k} \{2^{\mathrm{DF}} g^{p'',k}_{a,k} - {}^{\mathrm{DF}} g^{k,p''}_{a,k} \}, \tag{84}$$

where the summation runs over the active occupied orbitals. Each Coulomb integral has to be recovered from the precomputed and stored three-index quantities by contracting them over the auxiliary index Q. The final formulas can be written as

$$F^{val}_{ip''} = \sum_{kQ} \{2B_{Q,ip''}B_{Q,kk} - B_{Q,ik}B_{Q,p''k}\}, \tag{85}$$

$$F^{val}_{ap''} = \sum_{kQ} \{2B_{Q,ap''}B_{Q,kk} - B_{Q,ak}B_{Q,p''k}\}. \tag{86}$$

Until now all steps can be considered as the preliminary part of the CCSD(F12) calculation. The remaining, main part of the program, iteratively solves the CC equations. The solution of the CC equations requires the computation of the CC residuals (step 10; Alg. 1) that are defined by the projections onto the excitation manifolds [Eqs. (45), (46), (47)]. The values of the residuals allow for determining the next guess for the amplitudes and these new amplitudes (steps 12, 13; Alg. 1) lead to new residuals. The procedure is repeated until convergence is reached. The measure of the convergence is the difference between the energies in the iterations (k) and $(k+1)$ (step 11; Alg. 1) and the norm of the T_1 vector function. As the initial guess for the amplitudes, the optimal MP2 and MP2-F12 amplitudes are assumed for the conventional and explicitly correlated parts, respectively. To improve the convergence, the direct inversion in the iterative subspace (DIIS) method [52] is always used for the computation of the next guess for the amplitudes (step 13; Alg. 1).

4.4 The $V^{\mu,\nu}_{x,y}$ intermediate

Before discussing the formulas for the T_1, T_2 and $T_{2'}$ vector functions, it is important to introduce the $V^{\mu,\nu}_{x,y}$ intermediate. In this part of the thesis, detailed information about this quantity will be presented. The most important contributions to the CC residuals are recovered from $V^{\mu,\nu}_{x,y}$ and thus it can be considered as the "key intermediate" of the CCSD(F12) implementation in TURBOMOLE. In general, such quantity, in the MO basis, can be introduced in the following way

$$V^{r,s}_{p,q} = \langle pq | f_{12} \hat{Q}_{12} r^{-1}_{12} | rs \rangle, \tag{87}$$

where f_{12} is the correlation factor, $\hat{Q}_{12}$ is the strong orthogonality projector that defines Ansatz of the explicitly correlated wave function (see Subsection 4.2 for the details concerning f_{12} and $\hat{Q}_{12}$), and r^{-1}_{12} is the electron-electron Coulomb operator. The implementation of the CCSD (and thus CCSD(F12)) method in TURBOMOLE is, to a large extent, designed in an integral-direct fashion. It means that the most time-consuming quantities are computed in the AO-basis, combined together and later on transformed to the MO-basis. The implementation of the additional F12 terms (at least some of them) follows this idea and therefore the intermediate $V^{\mu,\nu}_{x,y}$ with two AO indices was introduced.

Among the F12 terms this is one of the most important quantities and will be discuss in detail here.
It is convenient to begin with the simplest case, *i.e.* the V intermediate derived within **An1** and the RHF reference, without the CABS contribution (A1-noCABS). If we take the general formula of the V intermediate [Eq. (87)] and substitute the expression of the strong orthogonality projector that defines Ansatz 1 [Eq. (61)] we will get the following equation

$$V^{r,s}_{p,q} = \langle pq|f_{12}((1-\hat{P}_1)(1-\hat{P}_2))r_{12}^{-1}|rs\rangle. \tag{88}$$

In the AO-based implementation of the CCSD(F12) model we need the $V^{\mu,\nu}_{x,y}$, where the xy are the geminal indices (efectively they belong to the occupied space within this implementation, see Subsection 3.2) and $\mu\nu$ belong to the covariant AO-basis. In the subsequent sections we will use the term covariant for the description of the quantities in the "ordinary" AO basis (denoted by upper-case "AO" left superscript) whereas the contravariant basis is used for the back-transformed quantities (*vide infra*) (denoted with lower-case "ao" left superscript). The final equation for the A1-noCABS V intermediate can be written in the following way

$$^{\text{A1-noCABS}}V^{\mu,\nu}_{x,y} = (fg)^{\mu,\nu}_{x,y} - \sum_{\gamma\delta} {}^{\text{ao}}f^{\gamma,\delta}_{x,y}\,{}^{\text{AO}}g^{\mu,\nu}_{\gamma,\delta}, \tag{89}$$

where $(fg)^{\mu,\nu}_{x,y}$ is the integral over the $(f_{12}r_{12}^{-1})$ operator

$$(fg)^{\mu,\nu}_{x,y} = \langle xy|f_{12}r_{12}^{-1}|\mu\nu\rangle. \tag{90}$$

The ${}^{\text{ao}}f^{\gamma,\delta}_{x,y}$ are the above mentioned back-transformed f_{12} integrals and ${}^{\text{AO}}g^{\mu,\nu}_{\gamma,\delta}$ are Coulomb integrals computed in the covariant AO-basis. The back-transformed quantities are obtained from the quantities in the MO-basis by contracting them with the AO-MO coefficients over the MO indices. The integrals in the MO-basis are available from the preceding MP2-F12 calculation and the AO-MO coefficients are always available as the solution of the Hartree-Fock equations. The ${}^{\text{ao}}f^{\gamma,\delta}_{x,y}$ is computed as follows

$${}^{\text{ao}}f^{\gamma,\delta}_{x,y} = \sum_{pq} C_{\gamma p} f^{p,q}_{x,y} (C_{\delta q})^{\dagger}. \tag{91}$$

Obviously, the contravariant AO-basis is not equivalent with the covariant basis because of the non-orthogonality of the AO basis functions. If we try to transform the contravariant quantity again to the MO basis with the AO-MO coefficients, we will get the following equation

$$\begin{aligned}\sum_{\gamma\delta}(C_{\gamma r})^{\dagger}({}^{\text{ao}}f^{\gamma,\delta}_{x,y})C_{\delta s} &= \sum_{\gamma\delta}(C_{\gamma r})^{\dagger}\left(\sum_{pq} C_{\gamma p} f^{p,q}_{x,y}(C_{\delta q})^{\dagger}\right)C_{\delta s} \\ &= \sum_{pq}\left(\sum_{\gamma}(C_{\gamma r})^{\dagger}C_{\gamma p}\right) f^{p,q}_{x,y}\left(\sum_{\delta}(C_{\delta q})^{\dagger}C_{\delta s}\right) = \mathbf{C}^{\dagger}\mathbf{C}\mathbf{f}_{xy}\mathbf{C}^{\dagger}\mathbf{C}.\end{aligned} \tag{92}$$

Comparison of the r.h.s. of Eq. (92) with the well known formula [53]

$$\mathbf{C}^{\dagger}\mathbf{S}\mathbf{C} = \mathbf{1} \tag{93}$$

shows that the co- and contravariant AO spaces are equivalent only if the overlap matrix **S** is the unity matrix, which is not the case.
The inclusion of the CABS changes significantly the way the V intermediate is computed, because the CABS molecular orbital involves the contributions from both the conventional and the CABS AO spaces

$$\phi_{p''} = \sum_{\mu} C_{\mu p''}\phi_{\mu} + \sum_{\mu''} C_{\mu'' p''}\phi_{\mu''}. \tag{94}$$

The formulas for the projectors in Eqs. (75) and (79) suggest that the A1-noCABS expression has to be supplemented with the terms that depend explicitly on the CABS basis. The expression for the A1/3-CABS V intermediate can be written as

$$^{\text{A1/3-CABS}}V^{\mu,\nu}_{x,y} = (fg)^{\mu,\nu}_{x,y} - \sum_{\gamma\delta} {}^{\text{ao}}f^{\gamma,\delta\text{AO}}_{x,y} g^{\mu,\nu}_{\gamma,\delta} - \hat{P}^{\mu\nu}_{xy} \sum_{\gamma''\delta} {}^{\text{ao}}f^{\gamma'',\delta\text{AO}}_{x,y} g^{\mu,\nu}_{\gamma'',\delta}. \tag{95}$$

The $(fg)^{\mu,\nu}_{x,y}$ integrals are, as before, the four-index integrals with the $(f_{12}r_{12}^{-1})$ operator [Eq. (90)]. The back-transformed integrals ${}^{\text{ao}}f^{\gamma,\delta}_{x,y}$ are more involved because, in addition to the pure orbital contribution Eq. (91), there is also the contribution that involves the CABS index

$$^{\text{ao/A1}}f^{\gamma,\delta}_{x,y} = \sum_{pq} C_{\gamma p} f^{p,q}_{x,y} (C_{\delta q})^{\dagger} + \hat{P}^{\gamma\delta}_{xy} \sum_{p''q} C_{\gamma p''} f^{p'',q}_{x,y} (C_{\delta q})^{\dagger}. \tag{96}$$

This equation refers to the strong orthogonality operator Eq. (75), where the projection space is spanned by the full computational space of MOs. In the case of **An3** the analogous space involves the full occupied space (active + inactive MOs), therefore the summation index q in terms on the r.h.s. of Eq. (96) is replaced with J, where J belongs to the full occupied space. The **An3** formula can be written as

$$^{\text{ao/A3}}f^{\gamma,\delta}_{x,y} = \sum_{pJ} C_{\gamma p} f^{p,J}_{x,y} (C_{\delta J})^{\dagger} + \hat{P}^{\gamma\delta}_{xy} \sum_{p''J} C_{\gamma p''} f^{p'',J}_{x,y} (C_{\delta J})^{\dagger}. \tag{97}$$

The operator $\hat{P}^{\gamma\delta}_{xy}$ used in Eqs. (95), (96), (97) is a permutation operator (called sometimes symmetrizer), the following equation illustrates its action on the arbitrary four-index quantity A^{rs}_{pq}

$$\hat{P}^{rs}_{pq} A^{rs}_{pq} = A^{rs}_{pq} + A^{sr}_{qp}. \tag{98}$$

The orbital/CABS and CABS/CABS AO-MO coefficients $C_{\gamma p''}$, $C_{\gamma'' p''}$ are obtained by orthogonalizing the CABS to the orbital basis with the assumption that the CABS orbitals are also mutually orthogonal

$$\langle p|q''\rangle = 0, \tag{99}$$
$$\langle p''|q''\rangle = \delta_{p''q''}. \tag{100}$$

These coefficients are available from the preceding MP2-F12 calculation (Table 1). The presence of the symmetrizer $\hat{P}^{\mu\nu}_{xy}$ in Eq. (95) comes from the fact that in Eq. (75) there are two symmetric contributions. For the RHF reference and spin diagonal components of the UHF reference only one of these contributions has to be computed explicitly, the other one is recovered with the use of the permutational symmetry of the integrals. The

$\alpha\beta$ component, in the case of the UHF reference wave function, is more complicated. It will be described in more detail in the next section where the issues associated with the UHF reference are discussed. The back-transformed integrals ${}^{ao}f^{\gamma'',\delta}_{x,y}$ in Eq. (95) are obtained from the analogous quantities in the MO basis (all f_{12} integrals in the MO basis are computed within the preceding MP2-F12 run, see Table 1) by contracting them with appropriate coefficients

$$ {}^{ao/A1}f^{\gamma'',\delta}_{x,y} = \sum_{p''q} C_{\gamma''p''} f^{p'',q}_{x,y} (C_{\delta q})^{\dagger}. \tag{101} $$

Similar to Eq. (96), for Ansatz 3 one replaces the summation index q with J

$$ {}^{ao/A3}f^{\gamma'',\delta}_{x,y} = \sum_{p''J} C_{\gamma''p''} f^{p'',J}_{x,y} (C_{\delta J})^{\dagger}. \tag{102} $$

The ${}^{AO}g^{\mu,\nu}_{\gamma'',\delta}$ integrals in Eq. (95) are computed in a complete analogy to the ordinary $g^{\mu,\nu}_{\gamma,\delta}$ integrals, simply one of the AO indices is contained in a different basis set.

The formulas and remarks referring to the back-transformation of the f_{12} integrals are associated with step 5 in Algorithm 1. The implementation of the back-transformation procedure is not complicated and is illustrated in Algorithm 2.

Data: integrals in the MO-basis $f^{p^{('')},q}_{x,y}$, AO-MO coefficients $C_{\mu p}, C_{\mu p''}, C_{\mu'' p''}$
Result: back-transformed integrals $f^{\delta^{('')},\gamma}_{x,y}$
for *each pair* (x, y) **do**
  Read the matrix $f_{p^{('')}q}$
  Compute the appropriate contributions to the final quantity

$$ {}^{ao}f_{\gamma\delta} \longleftarrow \sum_{pq} (C_{\gamma p})^{\dagger} f_{pq} C_{\delta q} $$
$$ {}^{ao}f_{\gamma\delta} \longleftarrow \sum_{p''q(J)} (C_{\gamma p''})^{\dagger} f_{p''q(J)} C_{\delta q(J)} $$
$$ {}^{ao}f_{\gamma''\delta} \longleftarrow \sum_{p''q(J)} (C_{\gamma''p''})^{\dagger} f_{p''q(J)} C_{\delta q(J)} $$

  Store the back-transformed matrices on file
end

Algorithm 2: The procedure that carries out the back-transformation of the f_{12} integrals.

Transformation of each index is implemented as matrix-matrix multiplication using the efficient BLAS [54] procedures. Depending on the contraction dimension q or J one obtains the formulas for Ansätze 1 or 3, respectively.

Once the back-transformed f_{12} integrals are computed, the next step is to obtain the V intermediate [Eq. (95)]. This operation is split into two parts, the computation of the contraction between the f_{12} and Coulomb integrals over the AO indices, and supplementing this contraction with the $f_{12}r_{12}^{-1}$ integrals in order to form the final intermediate. The implementation of the CCSD/CCSD(F12) models in TURBOMOLE is partially integral-direct, and precisely this contraction is designed in such a fashion [48]. The Algorithm 3 is a part of the conventional code [48], for the purpose of the present work it has been generalized to treat the CABS index. Obviously the quantities with CABS indices are not present in the implementation of conventional CCSD [48].

Data: back-transformed integrals $f^{\mu,\nu}_{x,y}$, $f^{\mu'',\nu}_{x,y}$
Result: part of the V intermediate: $V^{\kappa,\lambda}_{x,y} = \sum_{\mu^{('')}\nu} f^{\mu^{('')},\nu}_{x,y} g^{\kappa,\lambda}_{\mu,^{('')}\nu}$
initializations
1 **for** *batches of shells* (**N**) **do**
2   **for** *shells* μ *and subsequently* μ'' **do**
3   read the integrals $f^{\mu^{('')},\eta}_{x,y}$ for $\mu^{('')} \in \mathbf{N}$ and all η
4   **for** *shells* $\nu \leq \mu^{('')}$ **do**
5   compute the Coulomb integrals for shells $\nu, \mu^{('')}$
6   **for** *shells* $\kappa \leq \mu^{('')}$ **do**
7   **for** *shells* $\lambda \leq \kappa$ *(if* $(\kappa = \mu)$ $\lambda \leq \mu$*)* **do**
          if *no CABS* **then**
8   contraction 1: $V^{\kappa,\mu}_{x,y} = \sum_{\alpha\beta} f^{\lambda_\alpha,\nu_\beta}_{x,y} g^{\kappa,\mu}_{\lambda_\alpha,\nu_\beta}$
            if $\lambda < \kappa$ **then**
9   contraction 2: $V^{\lambda,\mu}_{x,y} = \sum_{\alpha\beta} f^{\kappa_\alpha,\nu_\beta}_{x,y} g^{\lambda,\mu}_{\kappa_\alpha,\nu_\beta}$
            end
          end
          if $(\nu < \mu^{('')}$ *and* $(\mu^{('')}\nu) > (\kappa\lambda))$ **then**
10   contraction 3: $V^{\kappa,\nu}_{x,y} = \sum_{\alpha\beta} f^{\lambda_\alpha,\mu^{('')}_\beta}_{x,y} g^{\kappa,\nu}_{\lambda_\alpha,\mu^{('')}_\beta}$
          end
          if $(\nu < \mu^{('')}$ *and* $\lambda < \kappa)$ **then**
11   contraction 4: $V^{\lambda,\nu}_{x,y} = \sum_{\alpha\beta} f^{\kappa_\alpha,\mu^{('')}_\beta}_{x,y} g^{\lambda,\nu}_{\kappa_\alpha,\mu^{('')}_\beta}$
          end
        end
      end
    end
    write the intermediate to file
  end
end

Algorithm 3: The algorithm of the contraction between the back-transformed f_{12} and Coulomb integrals.

The Algorithm 3 begins with the loop that batches the f_{12} integrals into pieces which fit into the available memory (step 1; Alg. 3). This is followed by the loop over the first AO index μ (step 2, Alg. 3,). The CABS AO indices are treated as the extension of the conventional atomic basis, therefore the loop over μ'' begins after the conventional indices μ. This is followed by reading the appropriate f_{12} integrals from file (step 3; Alg. 3). The next step is the beginning of the loop over the AO shell index ν (steps 4; Alg. 3). For particular pair of $(\mu^{('')}, \nu)$ indices, the AO Coulomb integrals are computed (step 5; Alg. 3) followed by the loops over the remaining two AO indices (steps 6, 7; Alg. 3). The inner part of the algorithm describes the contraction between the integrals over the basis functions that belong to a particular shell (with subscripts $\alpha\beta$). The contraction is always performed for a particular quadruple of the AO indices $(\mu^{('')}, \nu, \kappa, \lambda)$. To exploit the symmetry of the AO integrals, the whole contraction is split into four parts with

appropriate constraints imposed on the AO indices (steps 8, 9, 10, 11; Alg. 3). In order to make the algorithm transparent, the loops over the spins were skipped.
The contractions described above does not form the complete V intermediate. Compared to Eq. (95), the integral over the $f_{12}r_{12}^{-1}$ operator is missing. The form of this operator depends on the correlation factor and, in the case of linear r_{12}, simplifies to the overlap integral

$$\langle xy|r_{12}r_{12}^{-1}|pq\rangle = \langle xy|\hat{1}|pq\rangle = \delta_{xp}\delta_{yq}. \tag{103}$$

In the MO basis this is a product of Kronecker delta functions. In the case of the V intermediate needed in CCSD(F12) theory, the indices p and q are in the AO basis, and the overlap integrals are not the unity matrices anymore. Instead, they must be expressed as partially transformed overlap integrals

$$\langle xy|\mu\nu\rangle = S_{x\mu}S_{y\nu}. \tag{104}$$

For the correlation factor in the form of Eq. (60), the integrals are computed in the DF approximation by contracting the appropriate three-index quantities over the auxiliary index (step 2; Alg. 1)

$$(fg)_{x,y}^{\mu,\nu} = \sum_Q \left(G_{x\mu}^Q (\tilde{F}G)_{y\nu}^Q + (\tilde{F}G)_{x\mu}^Q G_{y\nu}^Q \right). \tag{105}$$

where $G_{x\mu}^Q$ is a three-index Coulomb integral and $(\tilde{F}G)_{y\nu}^Q$ is the intermediate that involves the integral over the $f_{12}r_{12}^{-1}$ operator. Both quantities are partially transformed to the MO basis. The computation of the four-index integrals Eq. (105) is part of the implementation of the CCSD(F12) method, but for practical reasons is carried out within the MP2-F12 module (see step 2 of Algorithm 1). The details about the implementation of the three-index quantities can be found elsewhere [51].
The remaining step forms the final V intermediate. This is done by simple subtracting from the integrals Eq. (105) the contraction between the f_{12} and Coulomb integrals, like in Eq. (95). The complete V intermediate is stored on file.

4.4.1 The T_1-dependent contribution to the V intermediate

For the evaluation of the $T_{2'}$ vector function it is necessary to compute the T_1-dependent contribution to the V intermediate. The inclusion of this quantity is required by the full formulation of the CCSD(F12) method with optimized $c_{i,j}^{x,y}$ coefficients present in the $T_{2'}$ operator [Eq. (40)]. For the approximations where these coefficients are not optimized (*e.g.* they may be kept frozen at prescribed values) the evaluation of the $T_{2'}$ residual is skipped, and thus the T_1-dependent contribution to V need not be considered. The issues concerning the alternative approximations to the CCSD(F12) model are discussed in more detail in Subsection 4.9.
The computation of the T_1-dependent contribution to the V intermediate is problematic, because the T_1-dependence involves the "internal" index of the intermediate, *i.e.* the index over which the integrals are contracted. This can be written in the following way

$$^{T1}V_{x,y}^{\mu,\nu} = -\hat{P}_{xy}^{\mu\nu} \sum_{Iq''} f_{x,y}^{I,q''} g_{I,q''}^{\mu,\nu}. \tag{106}$$

where the $\tilde{I}$ index refers to a T_1-transformed quantity. The T_1 transformation is simply the transformation of the particular quantity with the T_1 amplitudes, *e.g.*

$$A_{x,y}^{\tilde{i},j} = \sum_a t_i^a A_{x,y}^{a,j}, \tag{107}$$

where A is an arbitrary quantity and t_i^a are the T_1 amplitudes. The manner in which the T_1-dependence is embedded into the intermediate [Eq. (106)] requires the separate computation of the T_1-dependent and non T_1-dependent quantities. For the full CCSD(F12) approximation with optimized $c_{i,j}^{x,y}$ coefficients, the calculation of ${}^{\mathrm{T1}}V_{x,y}^{\mu,\nu}$ has to be repeated in each CC iteration. Since the V intermediate is one of the most expensive terms of the whole CCSD(F12) calculation this becomes quite soon the limiting step. Therefore, other approximations have been considered (see Subsection 4.9 for details). As it was shown before, the computation of V takes place partially in the AO space. The T_1-dependent V intermediate is computed in complete analogy to the standard V, but just before the back-transformation procedure the f_{12} integrals are transformed with the T_1 amplitudes

$${}^{\mathrm{ao/A3}}f_{x,y}^{\gamma'',\delta} = \sum_{p''J} C_{\gamma''p''} \left(\sum_b t_J^b f_{x,y}^{p'',b} \right) (C_{\delta J})^{\dagger} = \sum_{p''\tilde{J}} C_{\delta J} f_{x,y}^{\tilde{J},p''} (C_{\gamma''p''})^{\dagger}. \tag{108}$$

It is worth to mention that the dimensions of the T_1 transformation refers to the active electrons (included in the correlation treatment). On the other hand, due to the presence of the $\hat{O}_1$ and $\hat{O}_2$ projectors in the operator [Eqs. (76), (79)], the index J in Eq. (108) runs over the full occupied space (active+inactive occupied orbitals). Therefore, the index $\tilde{J}$ should be understood as the union of the occupied inactive ($\bar{J}$) and occupied T_1-transformed indices ($\tilde{j}$)

$$\{\tilde{J}\} = \{\bar{J}\} \cup \{\tilde{j}\}. \tag{109}$$

The T_1 dependence is included at the level of the back-transformation of the integrals, which is an early stage of the calculation of V. The inclusion of the T_1 dependence does not change the dimensions of the integrals Eq. (108), therefore the code could be organized in such a way that the same routines are exploited for the purpose of the calculation of both V intermediates.

4.4.2 UHF considerations

The implementation of the CCSD(F12) model in TURBOMOLE has been generalized for UHF reference wave functions and the algorithm described above can treat the cases where the α and β orbitals are not the same. In the case of an arbitrary four-index quantity, one can distinguish four independent spin contributions, $\alpha\alpha$, $\beta\beta$ (spin diagonal), $\alpha\beta$ and $\beta\alpha$ (spin off-diagonal). If all indices belong to the same basis set (*i.e.* all of them belong to the conventional orbital basis or all of them belong to the CABS), then the $\beta\alpha$ contributions can be generated from the $\alpha\beta$ block by the particle-exchange operator. If this is not the case (*e.g.* in the case of $f_{x,y}^{p'',q}$ integrals) then both $\alpha\beta$ and $\beta\alpha$ contributions have to be computed separately. Therefore, the computation of the V intermediate with the UHF wave function is more involved. In the case of the UHF wave function, the $C_{\mu p}^{\alpha}$ and $C_{\mu p}^{\beta}$

AO-MO coefficients have to be used for the back-transformation of the f_{12} integrals.

$$ {}^{\mathrm{ao/A3}}_{\mathrm{UHF}} f^{\gamma\alpha,\delta\alpha}_{x\alpha,y\alpha} = \sum_{pJ} C^{\alpha}_{\gamma p} f^{p\alpha,J\alpha}_{x\alpha,y\alpha} (C^{\alpha}_{\delta J})^{\dagger} + \hat{P}^{\gamma\delta}_{xy} \sum_{p''J} C^{\alpha}_{\gamma p''} f^{p''\alpha,J\alpha}_{x\alpha,y\alpha} (C^{\alpha}_{\delta J})^{\dagger}, \tag{110} $$

$$ {}^{\mathrm{ao/A3}}_{\mathrm{UHF}} f^{\gamma\beta,\delta\beta}_{x\beta,y\beta} = \sum_{pJ} C^{\beta}_{\gamma p} f^{p\beta,J\beta}_{x\beta,y\beta} (C^{\beta}_{\delta J})^{\dagger} + \hat{P}^{\gamma\delta}_{xy} \sum_{p''J} C^{\beta}_{\gamma p''} f^{p''\beta,J\beta}_{x\beta,y\beta} (C^{\beta}_{\delta J})^{\dagger}, \tag{111} $$

$$ {}^{\mathrm{ao/A3}}_{\mathrm{UHF}} f^{\gamma\alpha,\delta\beta}_{x\alpha,y\beta} = \sum_{pJ} C^{\alpha}_{\gamma p} f^{p\alpha,J\beta}_{x\alpha,y\beta} (C^{\beta}_{\delta J})^{\dagger} + \sum_{p''J} C^{\alpha}_{\gamma p''} f^{p''\alpha,J\beta}_{x\alpha,y\beta} (C^{\beta}_{\delta J})^{\dagger} + \sum_{Iq''} C^{\alpha}_{\gamma I} f^{I\alpha,q''\beta}_{x\alpha,y\beta} (C^{\beta}_{\delta q''})^{\dagger}. \tag{112} $$

Eqs. (110)–(112) are the UHF analogues of Eq. (97), equations referring to **An1** [Eq. (96)] can easily be obtained by changing the dimension of the contraction. It is important to note that the symmetrizer [Eq. (98)] is not present in the case of the spin off-diagonal quantity, each component of the ${}^{\mathrm{ao/A3}}_{\mathrm{UHF}} f^{\gamma\alpha,\delta\beta}_{x\alpha,y\beta}$ has to be computed explicitly. Eq. (102) in the UHF fashion can be written in the folowing way

$$ {}^{ao/A3}_{\mathrm{UHF}} f^{\gamma''\alpha,\delta\alpha}_{x\alpha,y\alpha} = \sum_{p''J} C^{\alpha}_{\gamma''p''} f^{p''\alpha,J\alpha}_{x\alpha,y\alpha} (C^{\alpha}_{\delta J})^{\dagger}, \tag{113} $$

$$ {}^{ao/A3}_{\mathrm{UHF}} f^{\gamma''\beta,\delta\beta}_{x\beta,y\beta} = \sum_{p''J} C^{\beta}_{\gamma''p''} f^{p''\beta,J\beta}_{x\beta,y\beta} (C^{\beta}_{\delta J})^{\dagger}, \tag{114} $$

$$ {}^{ao/A3}_{\mathrm{UHF}} f^{\gamma\alpha,\delta''\beta}_{x\alpha,y\beta} = \sum_{Iq''} C^{\alpha}_{\gamma I} f^{I\alpha,q''\beta}_{x\alpha,y\beta} (C^{\beta}_{\delta''q''})^{\dagger}, \tag{115} $$

$$ {}^{ao/A3}_{\mathrm{UHF}} f^{\gamma''\alpha,\delta\beta}_{x\alpha,y\beta} = \sum_{p''J} C^{\alpha}_{\gamma''p''} f^{p''\alpha,J\beta}_{x\alpha,y\beta} (C^{\beta}_{\delta J})^{\dagger}. \tag{116} $$

Due to the fact that the upper indices of the back-transformed integrals belong to different spaces of orbitals (ordinary orbital basis and CABS) both spin off-diagonal contributions have to be considered [Eqs. (115), (116)]. The UHF generalization of the procedure that computes the back-transformed integrals was conceptually straightforward, but technically it involves a number of issues. In short, the task here was to contract the f_{12} integrals in the MO basis with appropriate AO-MO coefficients. Due to the fact that the dimensions associated with the α and β spins are usually different, special attention had to be paid on these contractions. The back-transformed integrals shown above were later contracted with the Coulomb integrals using the scheme shown in Algorithm 3. Technically, the only major issue within this contraction was the handling of the $\alpha\beta$ and $\beta\alpha$ blocks of the integrals [Eqs. (115), (116)]. To avoid I/O operations and redundant computation of the AO Coulomb integrals, both components are read to the memory at the same time and contracted with the same set of the AO Coulomb integrals. The reason for using the same set of Coulomb integrals is that in the covariant AO basis ("pure" AO basis) there is no distinction between the AO indices that formally belong to different spins; such distinction appears when the AO quantity is transformed to the MO basis by the contraction with the spin-dependent AO-MO matrices. The remaining part of the UHF V intermediate, the integral over the $(f_{12}r_{12}^{-1})$ operator Eq. (105), is also computed separately for various spin cases. The final expressions for the V intermediate with the

UHF reference wave function are:

$$ {}^{\text{A1/3-CABS}}_{\text{UHF}}V^{\mu\alpha,\nu\alpha}_{x\alpha,y\alpha} = (fg)^{\mu\alpha,\nu\alpha}_{x\alpha,y\alpha} - \sum_{\gamma\delta}^{\text{ao}} f^{\gamma\alpha,\delta\alpha\text{AO}}_{x\alpha,y\alpha} g^{\mu\alpha,\nu\alpha}_{\gamma\alpha,\delta\alpha} - \hat{P}^{\mu\nu}_{xy} \sum_{\gamma''\delta}^{\text{ao}} f^{\gamma''\alpha,\delta\alpha\text{AO}}_{x\alpha,y\alpha} g^{\mu\alpha,\nu\alpha}_{\gamma''\alpha,\delta\alpha}, \tag{117} $$

$$ {}^{\text{A1/3-CABS}}_{\text{UHF}}V^{\mu\beta,\nu\beta}_{x\beta,y\beta} = (fg)^{\mu\beta,\nu\beta}_{x\beta,y\beta} - \sum_{\gamma\delta}^{\text{ao}} f^{\gamma\beta,\delta\beta\text{AO}}_{x\beta,y\beta} g^{\mu\beta,\nu\beta}_{\gamma\beta,\delta\beta} - \hat{P}^{\mu\nu}_{xy} \sum_{\gamma''\delta}^{\text{ao}} f^{\gamma''\beta,\delta\beta\text{AO}}_{x\beta,y\beta} g^{\mu\beta,\nu\beta}_{\gamma''\beta,\delta\beta}, \tag{118} $$

$$ \begin{aligned} {}^{\text{A1/3-CABS}}_{\text{UHF}}V^{\mu\alpha,\nu\beta}_{x\alpha,y\beta} = {} & (fg)^{\mu\alpha,\nu\beta}_{x\alpha,y\beta} - \sum_{\gamma\delta}^{\text{ao}} f^{\gamma\alpha,\delta\beta\text{AO}}_{x\alpha,y\beta} g^{\mu\alpha,\nu\beta}_{\gamma\alpha,\delta\beta} - \\ & \left(\sum_{\gamma\delta''}^{\text{ao}} f^{\gamma\alpha,\delta''\beta\text{AO}}_{x\alpha,y\beta} g^{\mu\alpha,\nu\beta}_{\gamma\alpha,\delta''\beta} + \sum_{\gamma''\delta}^{\text{ao}} f^{\gamma''\alpha,\delta\beta\text{AO}}_{x\alpha,y\beta} g^{\mu\alpha,\nu\beta}_{\gamma''\alpha,\delta\beta} \right). \end{aligned} \tag{119} $$

In the spin off-diagonal part of the V intermediate, the symmetrizer again could not be used, instead the explicit contraction was computed and accumulated on the final $\alpha\beta$ component of V. It is worth to note that Eqs. (117)–(119) are common for Ansätze 1 and 3, the difference is only at the level of the back-transformations where different dimensions of the contraction (referring to the full computational basis or full occupied space) are used.

The UHF generalization of the T_1-dependent contribution to the V intermediate affects only the computation of the back-transformed integrals [Eqs. (113)–(116)]. Similar to the RHF case [Eq. (108)] the appropriate index has to be transformed with the T_1 amplitudes. The only technical issue here was to use the T_1 amplitudes that belong to the appropriate spin. The T_1-dependent f_{12} integrals in the MO basis (before the back-transformation) from Eqs. (113)–(116) can be written in the following way

$$ f^{p''\alpha,\tilde{J}\alpha}_{k\alpha,l\alpha} = \sum_b {}^{\alpha}t^b_j f^{p''\alpha,b\alpha}_{k\alpha,l\alpha}, \tag{120} $$

$$ f^{p''\beta,J\beta}_{k\beta,l\beta} = \sum_b {}^{\beta}t^b_j f^{p''\beta,b\beta}_{k\beta,l\beta}, \tag{121} $$

$$ f^{I\alpha,q''\beta}_{k\alpha,l\beta} = \sum_a {}^{\alpha}t^a_i f^{a\alpha,q''\beta}_{k\alpha,l\beta}, \tag{122} $$

$$ f^{p''\alpha,J\beta}_{k\alpha,l\beta} = \sum_b {}^{\beta}t^b_j f^{p''\alpha,b\beta}_{k\alpha,l\beta}. \tag{123} $$

Similar to Eq. (108), the index with tilde refers to the appropriate sum of the subspaces [see Eq. (109)].

4.5 The energy equation

The general CCSD(F12) energy expression has the following form

$$ E = E_{\text{HF}} + \langle \text{HF}|[\hat{f}, T_1] + \tfrac{1}{2}[[\hat{\Phi}, \hat{T}_1], \hat{T}_1] + [\hat{\Phi}, \hat{T}_2 + T_{2'}]|\text{HF}\rangle, \tag{124} $$

where the term written with the red font refers to the new, explicitly correlated contribution. The Hamiltonian has been split,

$$ \hat{H} = \hat{f} + \hat{\Phi}, \tag{125} $$

into the Fock operator $\hat{f}$ and the fluctuation potential $\hat{\Phi}$ as suggested by Møller and Plesset [55]. An application of the Baker-Campbell-Hausdorff (BCH) expansion and the

Slater-Condon rules yields the following working expression [5]

$$E_{\mathrm{RCCSD}} = E_{\mathrm{RHF}} + \sum_{ia} F_{ia} t_i^a + \sum_{aibj} (t_{i,j}^{a,b} + t_i^a t_j^b) \left(2g_{i,j}^{a,b} - g_{j,i}^{a,b}\right) + \sum_{xiyj} c_{i,j}^{x,y} \left(2V_{x,y}^{i,j} - V_{y,x}^{i,j}\right) \tag{126}$$

for the RHF reference wave function. For the UHF reference the coupled-cluster energy equation can be written as

$$\begin{aligned} E_{\mathrm{UCCSD}} =& E_{\mathrm{UHF}} + \sum_{\sigma} \sum_{ia} F_{ia}^{\sigma} t_{i\sigma}^{a\sigma} + \\ & \frac{1}{2} \sum_{\sigma\sigma'} \sum_{aibj} (t_{i\sigma,j\sigma'}^{a\sigma,b\sigma'} + t_{i\sigma}^{a\sigma} t_{j\sigma'}^{b\sigma'}) \left(g_{i\sigma,j\sigma'}^{a\sigma,b\sigma'} - \delta_{\sigma\sigma'} g_{j\sigma,i\sigma'}^{a\sigma,b\sigma'}\right) + \\ & \frac{1}{2} \sum_{\sigma\sigma'} \sum_{xiyj} c_{i\sigma,j\sigma'}^{x\sigma,y\sigma'} \left(V_{x\sigma,y\sigma'}^{i\sigma,j\sigma'} - \delta_{\sigma\sigma'} V_{y\sigma,x\sigma'}^{i\sigma,j\sigma'}\right). \end{aligned} \tag{127}$$

In both energy equations the second term on the r.h.s. vanishes if one uses canonical Hartree-Fock orbitals (due to the Brillouin theorem), but formally these terms should be considered explicitly (*e.g.* in the case when localized orbitals are used). The last term is the explicitly correlated contribution that involves the $c_{i,j}^{x,y}$ coefficients and the $V_{x,y}^{i,j}$ intermediate. In the case of the full CCSD(F12) model the optimized $c_{i,j}^{x,y}$ coefficients are used for the energy evaluation. The energies of the simplified CC models are obtained by using the predetermined values of $c_{i,j}^{x,y}$ with an additional Lagrangian correction (see Subsection 4.9). $V_{x,y}^{i,j}$ is calculated from more general quantity $V_{x,y}^{\mu,\nu}$ discussed in detail in Subsection 4.4. In the RHF case [Eq. (126)], the transformation of $V_{x,y}^{\mu,\nu}$ to the occupied space was already shown [Eq. (82)], in the case of a UHF reference, the analogous transformation can be written in the following way

$$V_{x\sigma,y\sigma'}^{i\sigma,j\sigma'} = \sum_{\mu\nu} (C_{\mu i}^{\sigma})^{\dagger} V_{x\sigma,y\sigma'}^{\mu\sigma,\nu\sigma'} C_{\nu j}^{\sigma'}, \tag{128}$$

where the indices σ and σ' refer to various spin coordinates $\{\alpha, \beta\}$. The only technical issue to program this intermediate with the UHF reference was to use the AO-MO coefficients of the proper spin. The contraction between $c_{i,j}^{x,y}$ coefficients and V intermediate in both energy expressions [Eqs. (126), (127)] is computationally not a demanding operation because all the indices involved belong to the occupied space (which is usually small compared to the size of the virtual space). The geminal indices xy are equivalent to the occupied indices in the present implementation. The energy equation depends explicitly on the conventional amplitudes T_1 and T_2, the $c_{i,j}^{x,y}$ coefficients, the Coulomb integrals and the $V_{x,y}^{i,j}$ intermediate. The latter quantity is obtained in the preliminary part of the calculation (step 7; Alg. 1), stored and always available. The computation of the CCSD(F12) energy requires the determination of the conventional CC amplitudes and geminal $c_{i,j}^{x,y}$ coefficients. All these quantities are obtained by solving the coupled-cluster amplitude equations discussed in the following sections.

4.6 General remarks about amplitude equations

The CC equations are considered as solved when the set of the amplitudes involved in the CC treatment has been determined. One of the methods of solving the CC equations is the projection of the following function

$$\exp(-\hat{T})\hat{H}\exp(\hat{T})|\mathrm{HF}\rangle \tag{129}$$

onto various excitations manifolds [Eqs. (45), (46), (47)]. The number of such equations depends on the CC model, *e.g.* in the case of CCS we only have one amplitude equation for the singles, the CCSD model requires one more equation for the doubles and so on. The explicitly correlated contribution is obtained by supplementing the cluster operator by one more term $T_{2'}$ [see Eqs. (40), (42)]. This term depends on the $c_{i,j}^{x,y}$ coefficients that have to be determined together with the conventional amplitudes in an iterative manner. The values of the projections [Eqs. (45), (46), (47)] can be considered as the residuals of the nonlinear equations that have to be solved, the algorithm should find the CC amplitudes for which the residuals vanish.
In the case of the full CCSD(F12) model, to obtain the solution of the CC equations the T_1, T_2 conventional amplitudes and $c_{i,j}^{x,y}$ coefficients have to be determined. Compared to the conventional CCSD scheme, the explicit electron correlation requires an additional equation for the $c_{i,j}^{x,y}$ coefficients [see Eqs. (41), (47)]. The CCSD(F12) amplitude equations, obtained by projecting Eq. (129) onto the excitation manifolds, have the following form

$$0 = \langle\mu_1|\tilde{\hat{f}} + \tilde{\hat{\Phi}} + [\tilde{\hat{\Phi}}, \hat{T}_2 + \hat{T}_{2'}]|\mathrm{HF}\rangle, \tag{130}$$

$$0 = \langle\mu_2|\tilde{\hat{\Phi}} + [\tilde{\hat{f}}, \hat{T}_2 + \hat{T}_{2'}] + [\tilde{\hat{\Phi}}, \hat{T}_2 + \hat{T}_{2'}] + \frac{1}{2}[[\tilde{\hat{\Phi}}, \hat{T}_2], \hat{T}_2 + 2\hat{T}_{2'}]|\mathrm{HF}\rangle, \tag{131}$$

$$0 = \langle\mu_{2'}|\tilde{\hat{\Phi}} + [\tilde{\hat{f}}, \hat{T}_2 + \hat{T}_{2'}] + [\tilde{\hat{\Phi}}, \hat{T}_2]|\mathrm{HF}\rangle, \tag{132}$$

where the red font refers to new terms with respect to conventional CCSD. The $\hat{f}$ and $\hat{\Phi}$ are the Fock operator and the fluctuation potential, respectively [see Eq. (125)]. The tilde over the operator denotes the T_1 similarity-transformed quantity

$$\tilde{\hat{\Phi}} = \exp(-\hat{T}_1)\hat{\Phi}\exp(\hat{T}_1). \tag{133}$$

Eqs. (130)–(132), derived with the assumption of the generalized Brillouin condition [31], entirely define the CCSD(F12) model. It is worth to mention that this model is an approximation to the full CCSD-F12 method [41], where more explicitly correlated terms are taken into account. For instance the following commutators, quadratic in the $c_{i,j}^{x,y}$ coefficients or higher order in the fluctuation potential, are neglected in the CCSD(F12) scheme [41]

$$[[\tilde{\hat{\Phi}}, \hat{T}_{2'}], \hat{T}_{2'}], \quad [\tilde{\hat{\Phi}}, \hat{T}_{2'}], \quad [[\tilde{\hat{\Phi}}, \hat{T}_{2'}], \hat{T}_2]. \tag{134}$$

The first commutator has been neglected in in both Eqs. (131) and (132), whereas the remaining two commutators were neglected only in the $T_{2'}$ equation. The removal of entire commutators assures the eize-extensivity of the CCSD(F12) energy [5, 41]. The Eqs. (130)–(132) are of a general form that is not yet suitable for the implementation. In the present work very often the expressions "vector function" and "residual" are used. They always refer to the many-index quantity, defined by the right hand sites of these equations. The working expressions of the coupled-cluster T_1, T_2 and $T_{2'}$ residuals are discussed in next subsections.

4.7 The T_1 amplitude equation

The conventional T_1 amplitude equation is obtained from the expression Eq. (130), where $\langle\mu_1|$ is the manifold of singly excited determinants with respect to the Hartree-Fock state. In the case of the RHF reference such determinants on the *bra* side can be written as

$$\langle\mu_1| = \langle\text{RHF}|\hat{E}^{\dagger}_{ia} = \left\langle \begin{smallmatrix} a \\ i \end{smallmatrix} \right|, \tag{135}$$

where the excitation operator $\hat{E}_{ia}$ was defined in Eq. (54). Formally, the projection of the function Eq. (129) onto this manifold is a two-index quantity, because it can be evaluated for all possible singly excited determinants on the *bra* side. The careful derivation involving simple, but tedious algebra gives the following expressions for the T_1 vector function [5, 56]

$$\Omega_{ai} = \Omega^G_{ai} + \Omega^H_{ai} + \Omega^I_{ai} + \Omega^J_{ai}, \tag{136}$$

where the components are defined in the following way

$$\Omega^G_{ai} = 2\sum_{cdk} t^{c,d}_{i,k}\tilde{L}^{d,c}_{k,a}, \tag{137}$$

$$\Omega^H_{ai} = -2\sum_{kdl} t^{a,d}_{k,l}\tilde{L}^{d,i}_{l,k}, \tag{138}$$

$$\Omega^I_{ai} = 2\sum_{ck}(2t^{a,c}_{i,k} - t^{a,c}_{k,i})\tilde{F}_{kc}, \tag{139}$$

$$\Omega^J_{ai} = 2\tilde{F}_{ai}. \tag{140}$$

The G, H and I terms were obtained from the following projection

$$\Omega^G_{ai} + \Omega^H_{ai} + \Omega^i_{ai} = \left\langle \begin{smallmatrix} a \\ i \end{smallmatrix} \middle| [\tilde{\Phi}, T_2] \middle| \text{HF} \right\rangle, \tag{141}$$

whereas the J term was obtained from [see Eq. (130)]

$$\Omega^J_{ai} = \left\langle \begin{smallmatrix} a \\ i \end{smallmatrix} \middle| \tilde{f} \middle| \text{HF} \right\rangle. \tag{142}$$

$\tilde{L}_{pqrs}$ is a "2×Coulomb−Exchange" ("2C-K") combination of the T_1 similarity-transformed Coulomb integrals

$$\tilde{L}^{r,s}_{p,q} = 2\tilde{g}^{r,s}_{p,q} - \tilde{g}^{s,r}_{p,q}, \tag{143}$$

and $\tilde{F}_{ia}$ is a T_1 similarity-transformed Fock matrix

$$\tilde{F}_{pq} = \tilde{h}_{pq} + \sum_k \tilde{L}^{k,q}_{k,p}. \tag{144}$$

The T_1 similarity-transformed integrals in Eq. (143) are defined in the following way

$$\tilde{g}^{r,s}_{p,q} = \sum_{\kappa\lambda\mu\nu} g^{\mu,\nu}_{\kappa,\lambda}\Lambda^p_{\kappa p}\Lambda^p_{\lambda q}\Lambda^h_{\mu r}\Lambda^h_{\nu s}. \tag{145}$$

Lambda matrices contain the ordinary AO-MO transformation coefficients and the T_1-dependent part

$$\mathbf{\Lambda}^p = \mathbf{C}(\mathbf{I} - \mathbf{t}_1^T), \tag{146}$$
$$\mathbf{\Lambda}^h = \mathbf{C}(\mathbf{I} + \mathbf{t}_1), \tag{147}$$

where $\mathbf{C}$ is a matrix of AO-MO coefficients, $\mathbf{I}$ is a unity matrix and $\mathbf{t}_1$ is an auxiliary matrix defined in the following way

$$\mathbf{t}_1 = \begin{pmatrix} \mathbf{0} & \mathbf{0} \\ \{t_{ai}\} & \mathbf{0} \end{pmatrix}. \tag{148}$$

In the case of the CCSD(F12) model there are several additional contributions that have to be taken into account. In general, since the CABS basis is a natural extension of the orbital basis, the explicitly correlated contributions to the CC residuals can be split into two groups, pure orbital and CABS contributions. The latter one is present only in the case of **An3**. The derivation of the F12 contributions to the vector function can be easily done with the "correspondence principle". The main idea of this procedure is to associate the virtual indices of the conventional expressions with those of the complete basis

$$\{a, b, c, d, \ldots\} \longrightarrow \{\alpha, \beta, \gamma, \delta, \ldots\}. \tag{149}$$

The amplitudes carrying the indices of the complete basis have to be identified with the $c_{i,j}^{x,y}$ coefficients contracted with the f_{12} integrals

$$t_{i,j}^{c,d} \longrightarrow t_{i,j}^{\alpha,\beta} \equiv \sum_{xy} c_{i,j}^{x,y} f_{\alpha,\beta}^{x,y}, \tag{150}$$

where we introduce the geminal indices xy. Within the current implementation they belong to the occupied space ($\{x, y, \ldots\} \equiv \{k, l, \ldots\}$), but in general the space of geminal indices may be extended. This is useful, for instance, in the case of explicitly correlated coupled-cluster response theory [57]. Described above "correspondence principle" applied to Eq. (137) gives the following expression

$$\Omega_{ai}^{G,\mathrm{R12}} = \sum_{\alpha\beta k} (2t_{i,k}^{\alpha,\beta} - t_{k,i}^{\alpha,\beta}) \tilde{g}_{k,a}^{\beta,\alpha} = \sum_{\alpha\beta k} \sum_{xy} (2c_{i,k}^{x,y} - c_{k,i}^{x,y}) f_{\alpha,\beta}^{x,y} \tilde{g}_{k,a}^{\beta,\alpha}, \tag{151}$$

where the "2C-K" combination was carried out on the amplitudes rather than on the Coulomb integrals. The above contraction between the Coulomb and f_{12} integrals over the $\alpha\beta$ indices resembles the V intermediate discussed earlier (see Subsection 4.4). In fact, it is a special case of this quantity, and the final form of $\Omega_{ai}^{G,\mathrm{R12}}$ expressed through V can be written as

$$\Omega_{ai}^{G,\mathrm{R12}} = \sum_{xyk} (2c_{i,k}^{x,y} - c_{k,i}^{x,y})(V^\dagger)_{k,\tilde{a}}^{x,y}. \tag{152}$$

The T_1 dependence is here associated with the "outer" index of the intermediate. It means that this quantity can be easily obtained from the precomputed $V_{x,y}^{\mu,\nu}$ intermediate. As it was mentioned before, the $\mu\nu$ indices belong to the covariant AO basis, and they can be easily transformed to another arbitrary subspace of MOs with appropriate AO-MO coefficients or $\mathbf{\Lambda}$ matrices. In the case of Eq. (152), such transformation can be written in the following way

$$(V^\dagger)^{x,y}_{k,\tilde{a}} = \sum_{\mu\nu}(C_{\mu k})^\dagger (V^\dagger)^{x,y}_{\mu,\nu}\Lambda^h_{\nu a}, \tag{153}$$

where the $\Lambda^h_{\nu a}$ matrix is defined in Eq. (147).
Performing similar consideration for the remaining contributions to the T_1 vector function, one obtains the following formulas

$$\Omega^{H,\mathrm{R12}}_{ai} = -\sum_{q''kl}(2t^{a,q''}_{l,k} - t^{q'',a}_{l,k})\tilde{g}^{q'',i}_{k,l}, \tag{154}$$

$$\Omega^{I,\mathrm{R12}}_{ai} = \sum_{q''k}(2t^{a,q''}_{i,k} - t^{q'',a}_{i,k})\tilde{F}_{kq''}. \tag{155}$$

These two contributions can be combined together as

$$\Omega^{I+H,\mathrm{R12}}_{ai} = \sum_{q''kl}(2t^{a,q''}_{l,k} - t^{q'',a}_{l,k})(\tilde{F}_{kq''}\delta_{li} - \tilde{g}^{q'',i}_{k,l}). \tag{156}$$

The $\tilde{g}^{q'',i}_{k,l}$ are the T_1 similarity-transformed Coulomb integrals with one index in the CABS space. The amplitudes $t^{a,q''}_{i,j}$, called the F12 amplitudes, are defined in the following way

$$t^{a,q''}_{i,j} = \sum_{xy} c^{x,y}_{i,j} f^{a,q''}_{x,y}. \tag{157}$$

where $f^{a,q''}_{x,y}$ are two-electron integrals over the correlation factor. The correlation factor is an arbitrary function of the distance between the electrons. In the case of a linear r_{12} term the entire scheme becomes the CCSD(R12) model. The Fock matrix $\tilde{F}_{kq''}$ in Eqs. (154) and (156) is a T_1-dependent quantity obtained in the following way

$$\tilde{F}_{kq''} = \sum_{dl}\left(2g^{p'',d}_{k,l} - g^{d,p''}_{k,l}\right)t^l_d. \tag{158}$$

The expressions (152), (154) and (155) are the only additional terms that contribute to the T_1 vector function within the CCSD(F12) model. They have appeared because the cluster operator was supplemented with the additional term [Eq. (42)] and this term entered the T_1 residual [Eq. (130)]. The $\Omega^{G,\mathrm{R12}}_{ai}$ contribution is obtained from the V intermediate together with the conventional contributions to the singles vector function T_1. The $\Omega^{I,\mathrm{R12}}_{ai}$ and $\Omega^{H,\mathrm{R12}}_{ai}$ are calculated in the subroutine that accounts for the C and D contributions to the conventional doubles. Both issues will be described in the appropriate sections. Together with the conventional terms [Eqs. (137)–(140)], the presented expressions constitute the complete, explicitly correlated T_1 vector function.

4.7.1 UHF considerations

All expressions from the previous section can be applied to the cases where the RHF reference function is available. The presence of the "2C-K" combinations in the formulas comes from the sum over spin contributions. In the UHF formalism, the spin cannot be integrated out because the spatial parts of α and β orbitals are not the same, and instead of the "2C-K" combinations, "C-K" should be considered. Moreover, due to the fact that the spin function are orthonormal, the $\alpha\beta$ exchange integral vanishes

$$\langle p\sigma', q\sigma'' || r\sigma', s\sigma'' \rangle = \langle p\sigma', q\sigma'' | r\sigma', s\sigma'' \rangle - \delta_{\sigma'\sigma''} \langle p\sigma', q\sigma'' | s\sigma'', r\sigma' \rangle, \tag{159}$$

and effectively, for the spin off-diagonal contributions, only the Coulomb component should be computed. The T_1 vector function is formally a two-index quantity, and both indices belong to the same electron. The analogue of the RHF projection manifold [Eq. (135)] suitable for the UHF reference wave function depends explicitly on the spin indices

$$\langle \mu_1 | = \langle \mathrm{UHF} | (\hat{\tau}_{i\sigma}^{a\sigma})^\dagger = \left\langle \begin{matrix} a\sigma \\ i\sigma \end{matrix} \right| . \tag{160}$$

The spin index σ refers here to either α or β spin coordinate, therefore both spin cases have to be considered separately.
The $\Omega_{ai}^{G,\mathrm{R12}}$ contribution to the T_1 vector function, in the UHF formalism, can be written in the following way

$$^{\sigma}\Omega_{ai}^{G,\mathrm{R12}} = \sum_{xyk} \sum_{\sigma'} c_{i\sigma,k\sigma'}^{x\sigma,y\sigma'} (V^\dagger)_{k\sigma,\tilde{a}\sigma'}^{x\sigma,y\sigma'}, \tag{161}$$

where the summation σ' runs over α and β spins. It is important to mention that in Eq. (161) contributions belonging to different spins are mixed together to get the final quantity. The V intermediate in Eq. (161) is obtained from previously discussed, more general quantity $V_{x,y}^{\mu,\nu}$ (see Subsection 4.4.2). Similar to the RHF case [Eq. (153)] the proper AO-MO matrices of transformation (ordinary $\mathbf{C}$ or T_1-dependent $\mathbf{\Lambda}$) have to be used

$$(V^\dagger)_{k\sigma,\tilde{a}\sigma'}^{x\sigma,y\sigma'} = \sum_{\mu\nu} (C_{\mu k}^{\sigma})^\dagger (V^\dagger)_{\mu\sigma,\nu\sigma'}^{x\sigma,y\sigma'} \Lambda_{\nu a}^{\sigma',h}. \tag{162}$$

The AO-MO coefficients are now labeled with the spin index. The generalization of the code that transforms the V intermediate w.r.t. Eq. (162) was rather straightforward, the only technical issue was to keep proper dimensions when using AO-MO matrices of transformation for various spins. The UHF analogues of the $\Omega_{ai}^{I,\mathrm{R12}}$ and $\Omega_{ai}^{H,\mathrm{R12}}$ terms can be written in the following way

$$^{\sigma}\Omega_{ai}^{H,R12} = -\sum_{q''kl} \sum_{\sigma'} t_{l\sigma,k\sigma'}^{a\sigma,q''\sigma'} \tilde{g}_{k\sigma',l\sigma}^{q''\sigma',i\sigma}, \tag{163}$$

$$^{\sigma}\Omega_{ai}^{I,R12} = \sum_{q''k} \sum_{\sigma'} t_{i\sigma,k\sigma'}^{a\sigma,q''\sigma'} \tilde{F}_{kq''}^{\sigma'}, \tag{164}$$

with the following generalization of the F12 amplitudes

$$t_{i\sigma,j\sigma'}^{a\sigma,q''\sigma'} = \sum_{xy} c_{i\sigma,j\sigma'}^{x\sigma,y\sigma'} f_{x\sigma,y\sigma'}^{a\sigma,q''\sigma'}. \tag{165}$$

Due to the fact that the upper indices of $t_{i\sigma,j\sigma'}^{a\sigma,q''\sigma'}$ belong to different spaces of orbitals ($a \in$ MOs, $q'' \in$ CABS) the $\alpha\beta$ and $\beta\alpha$ spin cases had to be considered separately (see Subsection 4.4.2 for a similar discussion associated with the UHF generalization of the $V_{x,y}^{\mu,\nu}$ intermediate). This makes the calculations of this quantity with the UHF reference ca. 4 times more expensive, compared to the RHF case.

4.8 The T_2 and $T_{2'}$ amplitude equations

The T_2 and $T_{2'}$ vector functions can be obtained in the way that is analogous to the case of the T_1 residual. The function Eq. (129) is now projected onto the conventional doubles and geminal manifolds, Eqs. (131) and (132), where the following definition of the projection manifolds is used in the RHF formalism [5]

$$\langle\mu_2| = \langle \mathrm{RHF}|\tfrac{1}{3}(\hat{E}^\dagger_{ia}\hat{E}^\dagger_{jb} + \tfrac{1}{2}\hat{E}^\dagger_{ja}\hat{E}^\dagger_{ib}) = \left\langle \overline{\begin{matrix} ab \\ ij \end{matrix}} \right|, \tag{166}$$

$$\langle\mu_{2'}| = \sum_{\alpha\beta} w^{\alpha,\beta}_{x,y} \left\langle \overline{\begin{matrix} \alpha\beta \\ ij \end{matrix}} \right| = \left\langle \overline{\begin{matrix} xy \\ ij \end{matrix}} \right|. \tag{167}$$

In Eq. (167) the $\alpha\beta$ indices refer to the formally complete basis, and the coefficients $w^{\alpha,\beta}_{k,l}$ were defined in Eq. (59). The projection manifolds

$$\left\langle \overline{\begin{matrix} \alpha\beta \\ ij \end{matrix}} \right| \qquad \left\langle \overline{\begin{matrix} xy \\ ij \end{matrix}} \right| \tag{168}$$

in Eq. (167) are defined in analogy to Eq. (166). Let us start from the equations of conventional CCSD, the explicitly correlated contributions will be derived later with the "correspondence principle" discussed in the previous section. The complete T_2 vector function can be derived from Eq. (131) with the use of Wick's theorem, Slater-Condon rules or diagrammatic techniques [5, 56, 36]. After various algebraic manipulations this vector function can be decomposed into the following contributions [5, 56]

$$\Omega_{aibj} = \Omega^A_{aibj} + \Omega^B_{aibj} + 2\hat{P}^{ab}_{ij}\left(\Omega^C_{aibj} + \Omega^D_{aibj} + \Omega^E_{aibj}\right) + \Omega^F_{aibj}, \tag{169}$$

with the symmetrizer defined in Eq. (98), and the following definition of the contributions

$$\begin{aligned}
\Omega^A_{aibj} =& 2\sum_{kl} t^{a,b}_{k,l}\Big(\tilde{g}^{i,j}_{k,l} + \sum_{cd} t^{c,d}_{i,j} g^{c,d}_{k,l}\Big), && (170)\\
\Omega^B_{aibj} =& 2\sum_{cd} t^{c,d}_{i,j}\tilde{g}^{c,d}_{a,b}, && (171)\\
\Omega^C_{aibj} =& -(\tfrac{1}{2} + \hat{P}_{ij})\sum_{ck} t^{c,b}_{j,k}\Big(\tilde{g}^{i,c}_{k,a} - \tfrac{1}{2}\sum_{dl} t^{a,d}_{l,i} g^{d,c}_{k,l}\Big), && (172)\\
\Omega^D_{aibj} =& \tfrac{1}{2}\sum_{ck}(2t^{c,b}_{j,k} - t^{c,b}_{k,j})\Big(\tilde{L}^{i,c}_{a,k} - \tfrac{1}{2}\sum_{dl}(2t^{a,d}_{i,l} - t^{a,d}_{l,i})L^{d,c}_{l,k}\Big), && (173)\\
\Omega^E_{aibj} =& \sum_{c} t^{a,c}_{i,j}\Big(F_{b\tilde{c}} - \sum_{dlm} t^{d,b}_{l,m}L^{d,c}_{l,m}\Big) - \sum_{k} t^{a,b}_{i,k}\Big(F_{\tilde{k}j} - \sum_{dem} t^{d,e}_{j,m}L^{k,d}_{m,e}\Big) && (174)\\
\equiv& \Omega^{E1}_{aibj} + \Omega^{E2}_{aibj}, &&\\
\Omega^F_{aibj} =& 2\tilde{g}^{i,j}_{a,b}. && (175)
\end{aligned}$$

The operator $\hat{P}_{ij}$ interchanges the indices $i \leftrightarrow j$. These equations have been expressed through the T_1 similarity-transformed Coulomb integrals and Fock matrix Eqs. (145) and (144). The last term [Eq. 175] was obtained from the projection

$$\Omega^F_{aibj} = \left\langle \overline{\begin{matrix} ab \\ ij \end{matrix}} \middle| \tilde{\Phi} \middle| \mathrm{HF} \right\rangle. \tag{176}$$

The remaining terms that are linear in the T_2 amplitudes were obtained from the projection

$$\left\langle \overline{\begin{matrix} ab \\ ij \end{matrix}} \middle| [\tilde{\hat{f}} + \tilde{\hat{\Phi}}, \hat{T}_2] \middle| \mathrm{HF} \right\rangle, \tag{177}$$

whereas the terms square in the T_2 amplitudes come from the projection [see Eq. (131)]

$$\left\langle \overline{\begin{matrix} ab \\ ij \end{matrix}} \middle| [[\tilde{\hat{\Phi}}, \hat{T}_2], \hat{T}_2] \middle| \mathrm{HF} \right\rangle. \tag{178}$$

The F12 terms that contribute to the T_2 and $T_{2'}$ vector functions can be divided into three classes, depending on the upper indices of the conventional T_2 amplitudes. If both upper indices are present on the final vector function (ab), *e.g.* the first contribution in the term Ω^A_{aibj}

$$\sum_{kl} t^{a,b}_{k,l} \tilde{g}^{i,j}_{k,l}, \tag{179}$$

then the F12 contribution vanishes (due to the presence of the strong orthogonality projector $\hat{Q}_{12}$). If both indices are the virtuals not present on the final quantity (cd), *e.g.* the second contribution to the Ω^A_{aibj} term

$$\sum_{kl} t^{a,b}_{k,l} \sum_{cd} t^{c,d}_{i,j} g^{c,d}_{k,l}, \tag{180}$$

then the F12 analogue can be expressed through the V intermediate. Such terms do not vanish for both Ansätze 1 and 3. The remaining case is when only one index is a virtual belonging to the vector function (cb), *e.g.* the first contribution to the Ω^C_{aibj} term

$$\sum_{ck} t^{c,b}_{j,k} \tilde{g}^{i,c}_{k,a}. \tag{181}$$

The F12 analogues of such terms are present only for Ansatz 3. They are derived with the use of the F12 amplitudes [Eq. (157)].
Similar to the conventional case, the terms linear in the $c^{x,y}_{i,j}$ coefficients are obtained from the projection

$$\left\langle \overline{\begin{matrix} ab \\ ij \end{matrix}} \middle| [\tilde{\hat{f}} + \tilde{\hat{\Phi}}, \hat{T}_{2'}] \middle| \mathrm{HF} \right\rangle, \tag{182}$$

whereas the remaining F12 terms, linear in both $c^{x,y}_{i,j}$ coefficients and T_2 amplitudes, were obtained from the projection [see Eq. (131)]

$$\left\langle \overline{\begin{matrix} ab \\ ij \end{matrix}} \middle| [[\tilde{\hat{\Phi}}, \hat{T}_2], \hat{T}_{2'}] \middle| \mathrm{HF} \right\rangle. \tag{183}$$

From the conventional CCSD expressions [Eqs. (170)–(175)] one can expect three F12 contributions that can be derived and expressed in terms of the V intermediate. These are the analogues of the second contribution to the Ω^{A}_{aibj} term, the Ω^{B}_{aibj} term and the Ω^{E2}_{aibj} term. With simple but tedious algebra, one obtains the following formulas

$$\Omega^{A,\mathrm{R12}}_{aibj} = \sum_{klxy} t^{a,b}_{k,l} c^{x,y}_{i,j} (V^{\dagger})^{x,y}_{k,l}, \tag{184}$$

$$\Omega^{B,\mathrm{R12}}_{aibj} = \sum_{xy} c^{x,y}_{i,j} (V^{\dagger})^{x,y}_{\tilde{a},\tilde{b}}, \tag{185}$$

$$\Omega^{E2,\mathrm{R12}}_{aibj} = -\sum_{kl} t^{a,b}_{i,k} \sum_{xy} \left(2c^{x,y}_{l,j} - c^{x,y}_{j,l}\right) (V^{\dagger})^{x,y}_{l,k}. \tag{186}$$

All three contributions are obtained in one step by supplementing the conventional Ω^{BF}_{aibj} term with additional explicitly correlated contribution. The Ω^{BF}_{aibj} term combines the B and F terms [Eqs. (171), (175)] and has the following form in the MO basis [56]

$$\Omega^{BF}_{aibj} \equiv \Omega^{B}_{aibj} + \Omega^{F}_{aibj} = \sum_{cd} t^{c,d}_{i,j} \tilde{g}^{c,d}_{a,b} + \tilde{g}^{i,j}_{a,b}. \tag{187}$$

The contraction over the cd virtual indices in the above formula is the most expensive term within the whole conventional CCSD scheme. It scales like $n^2_{\mathrm{occ}} n^4_{\mathrm{vir}}$. Moreover, it requires the T_1 similarity-transformed Coulomb integrals with four virtual MO indices available. The storage of them on disk could very easily limit applicability of the code. Therefore, the AO-based integral-direct scheme was implemented here. The alternative approach to this transformation scales like $n^2_{\mathrm{occ}} N^4$ without carrying out expensive AO-MO transformation on the Coulomb integrals, where N is the number of basis functions (AO orbitals). There is an additional gain, because from the BF intermediate with two AO indices, one can compute the A and $E2$ contributions to the T_2 residual (*vide infra*). This intermediate, in the form that is appropriate for the AO-based algorithm, can be written in the following way

$$\Omega^{BF}_{\mu i\nu j} = \sum_{\kappa\lambda} \left(\sum_{cd} \Lambda^{h}_{\kappa c} \Lambda^{h}_{\lambda d} t^{c,d}_{i,j} + \Lambda^{h}_{\kappa i} \Lambda^{h}_{\lambda j} \right) g^{\kappa,\lambda}_{\mu,\nu} = \sum_{\kappa\lambda} M^{\kappa,\lambda}_{i,j} g^{\kappa,\lambda}_{\mu,\nu}. \tag{188}$$

where the $M^{\kappa,\lambda}_{i,j}$ density combines the $\mathbf{\Lambda}$ matrices and T_2 amplitudes

$$M^{\kappa,\lambda}_{i,j} = \sum_{cd} \Lambda^{h}_{\kappa c} \Lambda^{h}_{\lambda d} t^{c,d}_{i,j} + \Lambda^{h}_{\kappa i} \Lambda^{h}_{\lambda j}. \tag{189}$$

$\Omega^{BF}_{\mu i\nu j}$ is a "key quantity" among the conventional terms [48, 56]. The direct transformation of this intermediate to the MO basis immediately gives the Ω^{B}_{aibj} and Ω^{F}_{aibj} terms [Eqs. (171), (175)]

$$\sum_{\mu\nu} \Omega^{BF}_{\mu i\nu j} \Lambda^{p}_{\mu a} \Lambda^{p}_{\nu b} = \Omega^{BF}_{aibj} = \Omega^{B}_{aibj} + \Omega^{F}_{aibj}. \tag{190}$$

It is also used for the computation of the conventional Ω^{A}_{aibj} and Ω^{E2}_{aibj} terms [48]. They are obtained in the following way

$$\Omega^{A+E2}_{aibj} = \sum_{kl} t^{a,b}_{k,l} \Gamma'_{kilj}. \tag{191}$$

with the intermediates defined as

$$\Gamma'_{kilj} = \sum_{\mu\nu} \Omega^{BF}_{\mu i\nu j} \Lambda^{p}_{\mu k} \Lambda^{p}_{\nu l} - \delta_{ki} E^2_{lj} - \delta_{lj} E^2_{ki}, \tag{192}$$

$$E^2_{ij} = \sum_{\mu} \Lambda_{\mu i} G_{\mu j}, \tag{193}$$

$$G_{\mu i} = \sum_{\nu} F^{core}_{\mu\nu} \Lambda^{h}_{\nu i} + \sum_{\nu j} \Lambda^{p}_{\nu j} \left(2\Omega^{\mathrm{BF}}_{\mu i\nu j} - \Omega^{\mathrm{BF}}_{\nu i\mu j}\right). \tag{194}$$

The Fock density has been split here into the core and valence parts. The core contribution is included explicitly in the $G_{\mu i}$ term, the valence part is recovered from the $\Omega^{BF}_{\mu i\nu j}$ intermediate, both of them in Eq. (194). The $M^{\kappa,\lambda}_{i,j}$ density [Eq. (189)] is an important quantity in the implementation of the conventional terms. It is also used for the $T_{2'}$ residual, where problematic F12 terms can be recovered in a very elegant way (*vide infra*). The terms $\Omega^{A,\mathrm{R12}}_{aibj}$, $\Omega^{B,\mathrm{R12}}_{aibj}$ and $\Omega^{E2,\mathrm{R12}}_{aibj}$ [Eqs. (184)–(186)] are obtained by adding to the $\Omega^{BF}_{\mu i\nu j}$ intermediate [Eq. (188)] just one, not expensive contraction. The final explicitly correlated $\Omega^{BF,\mathrm{R12}}_{\mu i\nu j}$ intermediate can then be written as

$$\Omega^{BF,\mathrm{R12}}_{\mu i\nu j} = \sum_{\kappa\lambda} M^{\kappa,\lambda}_{i,j} g^{\kappa,\lambda}_{\mu,\nu} + \sum_{xy} c^{x,y}_{i,j} V^{\mu,\nu}_{x,y}, \tag{195}$$

where again the V intermediate was used here (discussed in Subsection 4.4). The explicit form of the terms in Eqs. (184)–(186) is calculated by proper transformation of the $\Omega^{BF,\mathrm{R12}}_{\mu i\nu j}$ intermediate [Eq. (195)] to the MO basis. There is full analogy between the conventional and F12 terms, therefore no additional effort in the case of the latter is required.
The F12 residual can be split into the following components in the MO basis

$$\Omega^{\mathrm{R12}}_{xiyj} = \Omega^{B,\mathrm{R12}}_{xiyj} + 2\hat{P}^{xy}_{ij}(\Omega^{C,\mathrm{R12}}_{xiyj} + \Omega^{D,\mathrm{R12}}_{xiyj} + \Omega^{E,\mathrm{R12}}_{xiyj}) + \Omega^{F,\mathrm{R12}}_{xiyj}. \tag{196}$$

with the following definition of the terms

$$\Omega^{B,\mathrm{R12}}_{xiyj} = \sum_{ab} t^{a,b}_{i,j} V^{\tilde{a},\tilde{b}}_{x,y}, \tag{197}$$

$$\Omega^{C,\mathrm{R12}}_{xiyj} = -\left(\tfrac{1}{2} + \hat{P}_{ij}\right) \sum_{p''b} f^{p'',b}_{x,y} \sum_{ck} t^{c,b}_{j,k} \tilde{g}^{k,p''}_{i,c}, \tag{198}$$

$$\Omega^{D,\mathrm{R12}}_{xiyj} = \tfrac{1}{2} \sum_{p''b} f^{p'',b}_{x,y} \sum_{ck} (2t^{b,c}_{j,k} - t^{c,b}_{j,k})(2\tilde{g}^{k,p''}_{c,i} - \tilde{g}^{k,p''}_{i,c}), \tag{199}$$

$$\begin{aligned}\Omega^{E,\mathrm{R12}}_{xiyj} &= \sum_{p''b} f^{p'',b}_{x,y} \sum_{c} t^{b,c}_{j,i} (F^{val}_{p''\tilde{c}} - F^{val}_{p''c}) + \sum_{mn} B^{m,n(ij)}_{x,y} c^{m,n}_{i,j} \\ &= \Omega^{E1,\mathrm{R12}}_{xiyj} + \Omega^{E',\mathrm{R12}}_{xiyj},\end{aligned} \tag{200}$$

$$\Omega^{F,\mathrm{R12}}_{xiyj} = V^{\tilde{i},\tilde{j}}_{x,y}. \tag{201}$$

We have neglected here discussed earlier commutators [Eq. (134)], what is an intrinsic feature of the CCSD(F12) model [41]. The $\Omega^{E',\mathrm{R12}}_{xiyj}$ and $\Omega^{F,\mathrm{R12}}_{xiyj}$ terms [Eqs. (200), (201)] were obtained from the projection

$$\Omega^{E',\mathrm{R12}}_{xiyj} + \Omega^{F,\mathrm{R12}}_{xiyj} = \left\langle \overline{\begin{matrix} xy \\ ij \end{matrix}} \middle| [\tilde{\hat{f}}, \hat{T}_{2'}] + \tilde{\hat{\Phi}} \middle| \mathrm{HF} \right\rangle, \tag{202}$$

and the remaining terms were derived from the projection [see Eq. (132)]

$$\Omega_{xiyj}^{B,\mathrm{R12}} + \Omega_{xiyj}^{C,\mathrm{R12}} + \Omega_{xiyj}^{D,\mathrm{R12}} + \Omega_{xiyj}^{E1,\mathrm{R12}} = \left\langle \overline{\begin{matrix} xy \\ ij \end{matrix}} \middle| [\tilde{\hat{\Phi}}, \hat{T}_{2'}] \middle| \mathrm{HF} \right\rangle. \tag{203}$$

The $\Omega_{xiyj}^{E',\mathrm{R12}}$ term [Eq. (200)] is computed explicitly by using the $B_{x,y}^{m,n(ij)}$ matrix from the preceding MP2-F12 calculation (see Table 1). This quantity has a complicated structure and it is considered as the most important term at the MP2-F12 level of theory. The mathematical form of $B_{x,y}^{m,n(ij)}$ can be written as

$$B_{x,y}^{m,n(ij)} = \langle xy|f_{12}\hat{Q}_{12}(\hat{f}_1 + \hat{f}_2 - \epsilon_i - \epsilon_j)\hat{Q}_{12}f_{12}|mn\rangle, \tag{204}$$

where ϵ_i, ϵ_j are the orbitals energies (for technical details referring to the implementation of $B_{x,y}^{m,n(ij)}$ see Ref. [49]). The contributions in Eqs. (197) and (201) are obtained in one step by contracting the $M_{i,j}^{\kappa,\lambda}$ density [Eq. (189)] with the $V_{x,y}^{\mu,\nu}$ intermediate (see Subsection 4.4) over the AO indices. This contraction can be written as

$$\begin{aligned} \sum_{\mu\nu} M_{i,j}^{\mu,\nu} V_{x,y}^{\mu,\nu} &= \sum_{\mu\nu} \Big(\sum_{ab} \Lambda_{\mu a}^{h} \Lambda_{\nu b}^{h} t_{i,j}^{a,b} + \Lambda_{\mu i}^{h} \Lambda_{\nu j}^{h} \Big) V_{x,y}^{\mu,\nu} = \sum_{ab} t_{i,j}^{a,b} V_{x,y}^{\tilde{a},\tilde{b}} + V_{x,y}^{\tilde{i},\tilde{j}} \\ &= \Omega_{xiyj}^{B,\mathrm{R12}} + \Omega_{xiyj}^{F,\mathrm{R12}}. \end{aligned} \tag{205}$$

The density $M_{i,j}^{\mu,\nu}$ is an expensive quantity, recomputed in each CC iteration. The main purpose of this quantity is to compute the $\Omega_{\mu i\nu j}^{BF}$ intermediate by contracting it with the AO Coulomb integrals [Eq. (188)]. Therefore this quantity is available anyway, independent of the F12 terms. This is also the reason why it is beneficial to use it in the F12 modules.

It is worth to mention that Eqs. (184)–(186) are the only F12 contributions to the conventional doubles with **An1**. The complete F12 part of the T_2 vector function in this case is calculated in a very efficient and elegant way. The additional costs are associated with the calculation of the V intermediate and with the contraction in Eq. (195). The $\Omega_{xiyj}^{B,\mathrm{R12}}$, $\Omega_{xiyj}^{E',\mathrm{R12}}$ and $\Omega_{xiyj}^{F,\mathrm{R12}}$ terms in Eqs. (197), (200) and (201) constitute the entire $T_{2'}$ residual within **An1**. The cost of the $\Omega_{xiyj}^{E',\mathrm{R12}}$ term in Eq. (200) is negligible, provided that the $B_{x,y}^{m,n(ij)}$ intermediate is available (see Table 1 for the details about the intermediates computed within preceding MP2-F12 calculation). The remaining two terms [Eqs. (197), (201)] are recovered in one step from the $M_{i,j}^{\kappa,\lambda}$ density and $V_{x,y}^{\mu,\nu}$ intermediate [Eq. (205)]. Here the only extra cost refers to the contraction [Eq. (205)] over the AO indices.

An3 requires several additional contributions that involve explicitly the CABS index. They are the analogues of the C, D and E^1 terms in the conventional case [48]. The E term is split into the Ω_{aibj}^{E1} and Ω_{aibj}^{E2} contributions [Eq. (174)] and the explicitly correlated analogue of the latter one has already been accounted for by using the V intermediate [Eqs. (191), (195)]. The remaining terms that have not been considered yet, *i.e.* Ω_{aibj}^{C}, Ω_{aibj}^{D} and Ω_{aibj}^{E1} [Eqs. (172), (173), (174)], have been grouped into one intermediate in the following way [48]

$$\Omega_{aibj}^{C+D+E1} = -\left(\tfrac{1}{2} + \hat{P}_{ij}\right) \sum_{ck} t_{j,k}^{c,b} C'_{ck,ai} + \tfrac{1}{2} \sum_{ck} (2t_{j,k}^{b,c} - t_{j,k}^{c,b}) D'_{ck,ai}, \tag{206}$$

with the intermediates defined as

$$C'_{ck,ai} = C_{ck,ai} - \tfrac{1}{2}\delta_{ik}E^{1}_{ac}, \tag{207}$$
$$D'_{ck,ai} = D_{ck,ai} + \tfrac{1}{2}\delta_{ik}E^{1}_{ac}, \tag{208}$$
$$C_{ck,ai} = \tilde{g}^{i,c}_{k,a} - \sum_{dl} t^{a,d}_{l,i}\, g^{c,d}_{l,k}, \tag{209}$$
$$D_{ck,ai} = \left(2\tilde{g}^{c,i}_{k,a} - \tilde{g}^{i,c}_{k,a}\right) + \sum_{dl}\left(2t^{a,d}_{i,l} - t^{d,a}_{i,l}\right)\left(2\tilde{g}^{d,c}_{l,k} - \tilde{g}^{c,d}_{l,k}\right), \tag{210}$$
$$E^{1}_{ac} = -F^{val}_{\tilde{a}c} + \tfrac{1}{4}\sum_{k}\left(3C_{ck,ak} - D_{ck,ak}\right), \tag{211}$$
$$F^{val}_{\tilde{a}c} = \sum_{k} t^{a}_{k} F^{val}_{kc}. \tag{212}$$

The explicitly correlated analogue of this contribution has the very similar structure

$$\begin{aligned}\Omega^{C+D+E1,\mathrm{R12}}_{aibj} =& -\left(\tfrac{1}{2} + \hat{P}_{ij}\right)\sum_{p''k} t^{p'',b}_{j,k} C'_{p''k,ai} + \tfrac{1}{2}\sum_{p''k}(2t^{b,p''}_{j,k} - t^{p'',b}_{j,k})D'_{p''k,ai}\\ &+\sum_{c} t^{a,c}_{i,j} E^{1'}_{bc},\end{aligned} \tag{213}$$

with similar definition of the intermediates

$$C'_{p''k,ai} = C_{p''k,ai} - \tfrac{1}{2}\delta_{ik}E^{1}_{ap''}, \tag{214}$$
$$D'_{p''k,ai} = D_{p''k,ai} + \tfrac{1}{2}\delta_{ik}E^{1}_{ap''}, \tag{215}$$
$$C_{p''k,ai} = \tilde{g}^{i,p''}_{k,a} - \sum_{dl} t^{a,d}_{l,i}\, g^{p'',d}_{l,k}, \tag{216}$$
$$D_{p''k,ai} = \left(2\tilde{g}^{p'',i}_{k,a} - \tilde{g}^{i,p''}_{k,a}\right) + \sum_{dl}\left(2t^{a,d}_{i,l} - t^{d,a}_{i,l}\right)\left(2\tilde{g}^{d,p''}_{l,k} - \tilde{g}^{p'',d}_{l,k}\right), \tag{217}$$
$$E^{1}_{ap''} = -F^{val}_{\tilde{a}p''} + \tfrac{1}{4}\sum_{k}\left(3C_{p''k,ak} - D_{p''k,ak}\right), \tag{218}$$
$$E^{1'}_{bc} = -\sum_{lmp''} t^{p'',b}_{l,m}\left(2g^{p'',c}_{l,m} - g^{c,p''}_{l,m}\right), \tag{219}$$
$$F^{val}_{\tilde{a}p''} = \sum_{k} t^{a}_{k} F^{val}_{kp''}. \tag{220}$$

In comparison to the conventional terms [Eqs. (206), (207)–(212)] one index has been moved to the CABS space ($c \rightarrow p''$) due to the application of the "correspondence principle" (see Subsection 4.7 for the discussion about this principle). The intermediate $E^{1'}_{bc}$ [Eq. (219)] has been also introduced as the additional F12 contribution to the E terms. To recover the Ω^{E1}_{aibj} contribution, one formally should carry out the explicit contractions as in Eq. (174). Within the implementation in TURBOMOLE, however, this contribution is obtained in a different way [48]. The contractions are carried out implicitly in Eqs. (206) and (213) by using modified (primed) intermediates. The traces of the C and D intermediates ($i = k$) are supplemented with the contributions from the E^1 intermediates [Eqs. (207), (208), (214), (215)], and when they are contracted with the amplitudes [Eqs. (206), (213)], the $\Omega^{E1,\mathrm{R12}}_{aibj}$ contribution to the residual is automatically recovered.

The terms that contribute to the F12 doubles have a similar structure

$$\Omega_{xiyj}^{C+D+E1,\mathrm{R12}} = \sum_{p''b} f_{x,y}^{p'',b} \left(-\left(\tfrac{1}{2} + \hat{P}_{ij}\right) \sum_{cm} t_{j,m}^{c,b} C'_{cm,p''i} + \tfrac{1}{2} \sum_{cm} (2t_{j,m}^{b,c} - t_{j,m}^{c,b}) D'_{cm,p''i} \right), \tag{221}$$

but there is an additional contraction with the f_{12} integrals which comes from the definition of the geminal projection manifold [Eq. (167)]. Due to the approximations made in the CCSD(F12) model, the definition of the intermediates is less involved

$$C'_{cm,p''i} = C_{cm,p''i} - \tfrac{1}{2}\delta_{mi} E^1_{p''c}, \tag{222}$$

$$D'_{cm,p''i} = D_{cm,tp''i} + \tfrac{1}{2}\delta_{mi} E^1_{p''c}, \tag{223}$$

$$C_{cm,p''i} = \tilde{g}_{m,i}^{p'',c}, \tag{224}$$

$$D_{cm,p''i} = 2\tilde{g}_{m,c}^{p'',i} - \tilde{g}_{m,c}^{i,p''}, \tag{225}$$

$$E^1_{p''c} = -F^{val}_{p''c} + \tfrac{1}{2} \sum_k (3C_{ck,p''k} - D_{ck,p''k}). \tag{226}$$

The explicitly correlated terms [Eqs. (213), (214)–(220), (221), (222)–(226)] are obtained within the routine that computes analogous conventional contributions [48] [Eqs. (206), (207)–(212)]. As it was already mentioned, the storage schemes of the quantities in TURBOMOLE are organized in such a way that the occupied indices (rather than virtual) are the fastest [48]. It means that during the contractions the virtuals are the "outer" indices, and this allows for a very elegant introduction of the CABS basis as the natural extension of conventional virtuals. In general, the computation of these terms is divided into two major steps. Within the first step the C'' and D' intermediates are prepared [Eqs. (214), (215), (222), (223)], in the second step the final contributions to the residuals are formed by contracting the intermediates with appropriate amplitudes and integrals [Eqs. (213), (221)]. Both steps are illustrated by schemes of the algorithms. Let us start with the Algorithms 4 and 5 where the implementation of the C'' and D' intermediates is presented.

In order to reuse the code, the existing routine that computes the conventional contributions [48] was generalized. The conventional and CABS C'' and D' intermediates are, to a large extent, computed in the same manner. Therefore, both cases are shortly discussed here. The algorithm starts with the initialization of the E intermediates with precomputed quantities (steps 1, 2; Alg. 4). Then, after reading the T_1 residual (step 3; Alg. 4), the T_1 amplitudes and three-index integrals (step 4; Alg. 4) the loop over batches of virtual (or CABS) orbitals begins (step 5; Alg. 4). Within this loop the T_2 amplitudes are read to the memory (step 6; Alg. 4) and proper combination of them is formed (step 7; Alg. 4). After reading the three-index quantities (step 8; Alg. 4) the Ω^I_{ai} and Ω^{II}_{ai} contributions to the T_1 residual (both conventional and explicitly correlated) are calculated (steps 9 and 10; Alg. 4, see Eqs. (138), (139), (154) and (155) for the discussion about these contributions). After accounting for the T_1-dependent contribution to the three-index integrals (step 11; Alg. 4) the loop over the virtual orbitals (followed by CABS) begins (step 12; Alg. 4), the Coulomb integrals are read to the memory and "2C-K" combination of them is formed (steps 13, 14; Alg. 4). The next step is to compute the T_1-dependent Coulomb integrals (step 15; Alg. 4) followed by the inner loop over the virtual index (again followed by CABS, step 16; Alg. 4). The inner part of the routine is presented on Algorithm 5.

The whole Algorithm 5 can be logically divided into two major parts, the calculation of the C'' intermediate and the analogous effort associated with the D' intermediate. The initialize the E intermediates:

$E^1_{bc} \leftarrow \sum_\mu \Lambda^p_{\mu b} F^{core}_{\mu c} + E^{1'}_{bc}$, $E^1_{ap''} \leftarrow \sum_i t_{ai} F^{val}_{ip''} - F^{val}_{ap''}$, $E_{p''c} \leftarrow F^{val}_{p''c}$

include the $C^{a,b}_{i,j}$ coupling matrix: $E^1_{ap''} \leftarrow F^{full}_{ap''}$, $E_{p''c} \leftarrow F^{full}_{p''c}$

Store E^1 intermediates on files

read the T_1 residual and $B_{Q,kl}$ integrals into the memory

read the amplitudes t^a_i and Fock matrices $\tilde{F}_{ia}$, $\tilde{F}_{ip''}$

for *(batches of virtuals A followed by batches of CABS orbitals P'')* **do**

  read the $t^{d,(vir)}_{l,i}$ amplitudes for all l, d, i and $vir \in A$ or $vir \in P''$

  form the "2C-K" combination: $\hat{t}^{d,(vir)}_{l,i} = 2t^{d,(vir)}_{l,i} - t^{d,(vir)}_{i,l}$

  read the three-index integrals: $B_{Q,i(vir)}$ for all Q, i and $vir \in A$ or $vir \in P''$

  for *(conventional virtuals ($vir \in A \equiv a$) or CABS orbitals ($vir \in P'' \equiv p''$))* **do**

    form the four-index quantities: $\tilde{g}^{k,a}_{l,i} = \sum_Q \hat{B}_{Q,kl} B_{Q,ia}$ or $\tilde{g}^{k,p''}_{l,i} = \sum_Q \hat{B}_{Q,kl} B_{Q,ip''}$

    compute the H contributions to the T_1 residual [Eqs. (138), (154)]:

    $\Omega^H_{dl} += -\sum_{ika} \hat{t}^{d,a}_{k,i} \tilde{g}^{l,a}_{k,i}$ and $\Omega^{H,\mathrm{R12}}_{dl} += -\sum_{ikp''} \hat{t}^{d,p''}_{k,i} \tilde{g}^{l,p''}_{k,i}$

    compute the I contributions to the T_1 residual [Eqs. (139), (155)]:

    $\Omega^I_{dl} += \sum_{ia} \hat{t}^{d,a}_{l,i} \tilde{F}_{ia}$ and $\Omega^{I,\mathrm{R12}}_{dl} += \sum_{ip''} \hat{t}^{d,p''}_{l,i} \tilde{F}_{ip''}$

  end

  read the three-index integrals: $\hat{B}_{Q,ai}$ for all Q, i and $a \in A$

  compute the T1-dependent contribution: $\hat{B}_{Q,ip''} \leftarrow B_{Q,ip''} + \sum_c t_{ci} B_{Q,cp''}$

  for *(conventional virtuals ($vir \in C \equiv c$) or CABS orbitals ($vir \in Q'' \equiv q''$))* **do**

    read the four-index integrals: $g^{d,c\in C}_{l,k}$ or $g^{d,q''\in Q''}_{l,k}$

    form the "2C-K" combination: $\hat{I}^{d,c}_{l,k} = 2g^{d,c}_{l,k} - g^{d,c}_{k,l}$

    read $B_{Q,kc}$ or $B_{Q,kq''}$ and E_{ac} or $E_{aq''}$ or $E_{p''c}$ for fixed c (or q'') from file

    read $B_{Q,ac}$ or $B_{Q,aq''}$ or $B_{Q,p''c}$ and add the T_1-dependent contribution:

    $\hat{B}_{Q,ac} \leftarrow B_{Q,ac} - \sum_k t_{ak} B_{Q,kc}$, $\hat{B}_{Q,aq''} \leftarrow B_{Q,aq''} - \sum_k t_{ak} B_{Q,kq''}$

    Compute the four-index integrals: $\tilde{g}^{i,c}_{k,a} = \sum_Q \hat{B}_{Q,ki} \hat{B}_{Q,ac}$,

    $\tilde{g}^{i,a}_{k,q''} = \sum_Q \hat{B}_{Q,ki} \hat{B}_{Q,q''a}$ or $\tilde{g}^{i,q''}_{k,a} = \sum_Q \hat{B}_{Q,ki} \hat{B}_{Q,aq''}$

    for *(conv. virtuals ($vir \in A \equiv a$) or CABS orbs. ($vir \in P'' \equiv p''$))* **do**

      the case with both CABS indices $p''q''$ is avoided

      see Algorithm 5 for the remaining part

    end

  end

end

Algorithm 4: Algorithm for the C', D' and E^1 intermediates (part I).

for *(batches of virtuals A followed by batches of CABS orbitals P'')* **do**

  for *(conventional virtuals ($vir \in C \equiv c$) or CABS orbitals ($vir \in Q'' \equiv q''$))* **do**

    for *(conv. virtuals ($vir \in A \equiv a$) or CABS orbs. ($vir \in P'' \equiv p''$))* **do**

      compute the T_2-dependent contribution to the C intermediate:

      $C^{c,a}_{k,i} \leftarrow -\frac{1}{2}\sum_{dl} t^{d,a}_{l,i} I^{d,c}_{l,k} \qquad C^{q'',a}_{k,i} \leftarrow -\sum_{dl} t^{d,a}_{l,i} I^{d,q''}_{l,k}$

      update the E intermediate:

      $E^1_{ac} += \frac{3}{2}\sum_k g^{k,c}_{k,a} + \frac{3}{2}\sum_k C^{c,a}_{k,k}$,

      $E^1_{aq''} += \frac{3}{2}\sum_k g^{k,q''}_{k,a} + \frac{3}{2}\sum_k C^{q'',a}_{k,k}$,

      $E^1_{cp''} += \frac{3}{2}\sum_k g^{k,c}_{k,p''}$

      update the C intermediate:

      $C^{c,a}_{k,i} += \tilde{g}^{i,c}_{k,a} \qquad C^{q'',a}_{k,i} += \tilde{g}^{i,q''}_{k,a} \qquad C^{c,p''}_{k,i} \leftarrow \tilde{g}^{i,c}_{k,p''}$

      compute the T_2-dependent contribution to the D intermediate:

      $D^{c,a}_{k,i} \leftarrow \frac{1}{2}\sum_{dl} t^{d,a}_{l,i} I^{d,c}_{l,k} \qquad D^{q'',a}_{k,i} \leftarrow \sum_{dl} t^{d,a}_{l,i} I^{d,q''}_{l,k}$

      compute the four-index integrals:

      $J^{c,a}_{k,i} \leftarrow -\frac{1}{2}\tilde{g}^{i,c}_{k,a} + \sum_Q B_{Q,kc}\hat{B}_{Q,ai}$,

      $J^{q'',a}_{k,i} \leftarrow -\frac{1}{2}\tilde{g}^{i,q''}_{k,a} + \sum_Q B_{Q,kq''}\hat{B}_{Q,ai}$,

      $J^{c,p''}_{k,i} \leftarrow -\frac{1}{2}\tilde{g}^{i,c}_{k,p''} + \sum_Q B_{Q,kc}\hat{B}_{Q,p''i}$

      update the E intermediate:

      $E^1_{ac} += -\sum_k J^{c,a}_{k,i} - \frac{1}{2}\sum_k D^{c,a}_{k,k}$,

      $E^1_{aq''} += -\sum_k J^{q'',a}_{k,k} - \frac{1}{2}\sum_k D^{q'',a}_{k,k}$,

      $E^1_{p''c} += -\sum_k J^{c,p''}_{k,k}$

      update the D intermediate:

      $D^{c,a}_{k,i} += 2J^{c,a}_{k,i} \qquad D^{q'',a}_{k,i} += 2J^{q'',a}_{k,i} \qquad D^{c,q''}_{k,i} \leftarrow 2J^{c,p''}_{k,i}$

      update the traces of C and D [form the final intermediates Eqs. (214), (215), (222), (223)]:

      $C^{c,a}_{k,k} \leftarrow C^{c,a}_{k,k} - \frac{1}{2}E^1_{ac} \qquad D^{c,a}_{k,k} \leftarrow D^{c,a}_{k,k} + \frac{1}{2}E^1_{ac}$

      $C^{q'',a}_{k,k} \leftarrow C^{q'',a}_{k,k} - \frac{1}{2}E^1_{aq''} \qquad D^{q'',a}_{k,k} \leftarrow D^{q'',a}_{k,k} + \frac{1}{2}E^1_{aq''}$

      $C^{c,p''}_{k,k} \leftarrow C^{c,p''}_{k,k} - \frac{1}{2}E^1_{p''c} \qquad D^{c,p''}_{k,k} \leftarrow D^{c,p''}_{k,k} + \frac{1}{2}E^1_{p''c}$

      save the intermediates C and D on files

    end

    save the intermediates E^1 on files

  end

end

save the T_1 residual Ω_{dl} on file

Algorithm 5: Algorithm for the C', D' and E^1 intermediates (part II).

first step is the computation of the T_2-dependent contribution to the C intermediate (step 1; Alg. 5) followed by the update of the E^1 intermediates (step 2; Alg. 5). The next contribution is accumulated on the C intermediate (step 3; Alg. 5) and the T_2-dependent contribution to the D intermediate is calculated (step 4; Alg. 5). The T_1 similarity-transformed Coulomb integrals are prepared (step 5; Alg. 5) and the second update of the E^1 intermediate is performed (step 6; Alg. 5). The next two steps are associated with the update of the D intermediate with the integrals (step 7; Alg. 5) and the update of the traces of C and D (step 8; Alg. 5) that forms the final intermediates (C' and D'). All intermediates as well as the T_1 residual are stored on files (steps 9, 10, 11; Alg. 5).
The final C, D and E^1 contributions to the T_2 and $T_{2'}$ amplitude equations require the contraction of the C' and D' intermediates with the amplitudes and integrals [Eqs. (213), (221)]. The implementation of these contractions is explained in Algorithm 6.

Data: C' and D' intermediates, amplitudes
Result: Final contributions to the T_2 and $T_{2'}$ residuals
1 read the current $T_{2'}$ residual Ω_{kilj} (so far there are E, F and G contributions [Eqs. (200), (201) and (197)])
2 **for** *(batches B of virtual b)* **do**
3   **for** *(spin cases and orbital types (virtual or CABS) for index c or q'')* **do**
4   read the amplitudes $t^{c,b}_{k,j}$ or $t^{q'',b}_{k,j}$ for all k, j, c (or q'') and $b \in B$
5   **for** *(spin cases and orbital types (virtual or CABS) for index a or p'')* **do**
      the case with both CABS indices ($p''q''$) is avoided
6   read the intermediates: $C'^{c,a}_{k,i}$, $D'^{c,a}_{k,i}$, $C'^{q'',a}_{k,i}$, $D'^{q'',a}_{k,i}$, $C'^{c,p''}_{k,i}$, $D'^{c,p''}_{k,i}$
7   for a: read the T_2 residual Ω_{aibj} for all i and j
      for p'': initialize $\Omega_{p''ibj} = 0$ for all i and j
      transpose the amplitudes: $\hat{t}^{c,b}_{k,j} = t^{c,b}_{j,k}$, $\hat{t}^{q'',b}_{k,j} = t^{q'',b}_{j,k}$
8   contract the amplitudes with the C' intermediate:
      $\tilde{\Omega}_{aibj} = \sum_{ck} t^{c,b}_{j,k} C'^{c,a}_{k,i}$ $\tilde{\Omega}_{p''ibj} = \sum_{ck} t^{c,b}_{j,k} C'^{c,p''}_{k,i}$ $\tilde{\Omega}_{aibj} = \sum_{q''k} t^{q'',b}_{j,k} C'^{q'',a}_{k,i}$
9   for a: build $\Omega_{aibj} = -\frac{1}{2}\left(\tilde{\Omega}_{aibj} + 2\tilde{\Omega}_{ajbi}\right)$
10   for p'': build $\Omega_{p''ibj} = -\frac{1}{2}\left(\tilde{\Omega}_{p''ibj} + 2\tilde{\Omega}_{p''jbi}\right)$
11   form the proper combination of the amplitudes:
      $\hat{t}^{c,b}_{k,j} = 2t^{c,b}_{k,j} - t^{b,c}_{k,j}$ or $\hat{t}^{q'',b}_{k,j} = 2t^{q'',b}_{k,j} - t^{b,q''}_{k,j}$
12   contract the amplitudes with the D' intermediate:
      $\Omega_{aibj} += \sum_{ck} t^{c,b}_{j,k} D'^{c,a}_{k,i}$ $\Omega_{p''ibj} += \sum_{ck} t^{c,b}_{j,k} D'^{c,p''}_{k,i}$
      $\Omega_{aibj} += \sum_{q''k} t^{q'',b}_{j,k} D'^{q'',a}_{k,i}$
      restore the amplitudes (undo step 11)
13   for a: write the residual Ω_{aibj} back to file (for all i and j)
14   for p'': update the residual $\Omega_{xiyj} += \sum_{p''b} f^{b,p''}_{y,x} \Omega_{p''ibj}$
    end
  end
end
15 write final residual Ω_{xiyj} to file

Algorithm 6: Algorithm for the contractions of the C' and D' intermediates with amplitudes.

The routine starts with reading the $T_{2'}$ residual (step 1; Alg. 6). This is because the $\Omega^{B,\mathrm{R12}}_{xiyj}$,

$\Omega^{E',\mathrm{R12}}_{xiyj}$ and $\Omega^{F,\mathrm{R12}}_{xiyj}$ contributions are already computed and stored on file [Eqs. (197), (200), (201)]. Then we have the loop that batches the virtual index b (step 2; Alg. 6) followed by the loop over the orbital indices c or CABS index q'' (step 3; Alg. 6). At the next step, the amplitudes are read to the memory (step 4; Alg. 6) and the inner loop over the virtual (or CABS) index a (or p'') begins (step 5; Alg. 6). In the inner part of the routine the intermediates are read to the memory (step 6; Alg. 6) and, for the case of the conventional terms, already computed contributions are also read to the memory (step 7; Alg. 6). For three combinations of CABS and orbital indices, the contractions of the C'' intermediate with the amplitudes are obtained (step 8; Alg. 6) and proper combinations of these contractions are formed (step 9, 10; Alg. 6). At step 11 appropriate combinations of conventional and F12 amplitudes are formed. The analogous contractions are obtained with the D' intermediate (step 12; Alg. 6) and the final T_2 residual is written back to file (step 13; Alg. 6). The calculation of the final $T_{2'}$ residual requires one more contraction due to the definition of the geminal projection manifold. In step 14, the intermediate $\Omega_{p''ibj}$ is contracted with f_{12} integrals. The last step of the routine stores the final Ω_{xiyj} residual on file (step 15; Alg. 6). In order to make Algorithms 4, 5 and 6 more transparent the loops over the spin indices have been omitted.

As it was already mentioned, the CABS indices are treated as the extension of conventional virtuals. Practically an additional outer loop over the basis types is introduced. This is illustrated in Algorithm 7.

mxicabs ⟵ 1
if *CABS is requested* **then** *mxicabs* ⟵ 2
for *(icabs = 1* **to** *mxicabs)* **do**
  if *(icabs = 1)* **then** orbitaltype ⟵ conventional virtual
  if *(icabs = 2)* **then** orbitaltype ⟵ CABS orbital
  if *(orbitaltype refers to conventional virtual)* **then**
1   Set up the range of the loop *(istart, istop)*
  else if *(orbitaltype refers to CABS)* **then**
2   Set up the range of the loop *(istart, istop)*
  end
  for *(iorbital = istart* **to** *istop)* **do**   /* set up the final loop */
    Carry out appropriate contractions with the quantities
  end
end

Algorithm 7: The introduction of the CABS to the routines.

There are several additional technical details, *e.g.* the ranges of the final loops have to be appropriately stored in the arrays (steps 1, 2; Alg. 7), but the general principle is very simple. Introducing CABS indices to the code in such a manner was possible, because the contraction dimension (over the occupied indices) is common for conventional and CABS contributions.

4.8.1 The coupling matrices

In addition to the terms presented so far, the implementation of the CCSD(F12) model in TURBOMOLE requires the inclusion of terms that are the analogues of the MP2-F12 terms responsible for the coupling between conventional and MP2-F12 amplitudes [47].

These terms enter the T_2 and $T_{2'}$ residuals in the following way [see Eqs. (131), (132)]

$$\Omega^{E0}_{aibj} = \left\langle \overline{\begin{matrix} ab \\ ij \end{matrix}} \right| [\hat{f}, \hat{T}_{2'}] |\text{HF}\rangle = \sum_{xy} C^{a,b}_{x,y} c^{x,y}_{i,j}, \tag{227}$$

$$\Omega^{E0}_{xiyj} = \left\langle \overline{\begin{matrix} xy \\ ij \end{matrix}} \right| [\hat{f}, \hat{T}_{2}] |\text{HF}\rangle = \sum_{ab} C^{a,b}_{x,y} t^{a,b}_{i,j}. \tag{228}$$

The MP2-F12 coupling matrix $C^{a,b}_{x,y}$ was obtained without the assumption of the generalized Brillouin condition [49]

$$C^{a,b}_{x,y} = \sum_{q''} f^{a,q''}_{x,y} F_{q''b} + \sum_{p''} f^{p'',b}_{x,y} F_{ap''}. \tag{229}$$

Within the present implementation of the CCSD(F12) model these terms are accounted for by supplementing the E^1 intermediates [Eqs. (218), (226)] with the Fock matrix elements

$$E^1_{ap''} \longleftarrow F^{full}_{ap''}, \tag{230}$$

$$E^1_{p''c} \longleftarrow F^{full}_{p''c}, \tag{231}$$

where the superscript $full$ denotes that the entire Fock density is included (one-electron, two-electron, core and valence). The implementation is embedded in the routine that computes the C', D' and E^1 intermediates. Step 2 in Algorithm 4 reflects this part of the code.

4.8.2 UHF considerations

All above equations have been derived and further implemented for the case of the UHF reference wave function. Many aspects of the UHF generalization have already been discussed in the subsection about the implementation of the T_1 residual (see Subsection 4.7.1). Here we will focus on some issues that are specific for the T_2 and $T_{2'}$ residuals. In the spin-orbital formalism (suitable when the UHF reference is assumed) the projection manifold can be defined in the following way

$$\langle \mu_2 | = \langle \text{UHF} | (\hat{\tau}^{a\sigma,b\sigma'}_{i\sigma,j\sigma'})^\dagger = \left\langle \begin{matrix} a\sigma, b\sigma' \\ i\sigma, j\sigma' \end{matrix} \right|, \tag{232}$$

$$\langle \mu_{2'} | = \sum_{\alpha\beta} w^{\alpha\sigma,\beta\sigma'}_{x\sigma,y\sigma'} \left\langle \begin{matrix} \alpha\sigma, \beta\sigma' \\ i\sigma, j\sigma' \end{matrix} \right| = \left\langle \begin{matrix} x\sigma, y\sigma' \\ i\sigma, j\sigma' \end{matrix} \right|, \tag{233}$$

where the *bra*'s in Eq. (233) are defined in analogy to Eq. (232). The UHF analogue of the coefficients needed for the representation of the geminal functions has the following form

$$w^{\alpha\sigma,\beta\sigma'}_{x\sigma,y\sigma'} = \langle \alpha\sigma, \beta\sigma' | \hat{Q}_{12} f_{12} | x\sigma, y\sigma' \rangle. \tag{234}$$

The terms of the T_2 residual that do not involve the explicit contraction over the CABS index [Eqs. (184), (185) and (186)] are obtained by supplementing the $\Omega^{BF}_{\mu i\nu j}$ intermediate with an additional term [Eq. (195)]. This additional contribution (second term on the r.h.s. of Eq. (195) in the RHF formalism), can be written in the following way using the spin-orbital notation

$$^{\sigma\sigma'}\Omega^{BF,\text{onlyR12}}_{\mu i \nu j} = \sum_{kl} c^{k\sigma,l\sigma'}_{i\sigma,j\sigma'} V^{\mu\sigma,\nu\sigma'}_{k\sigma,l\sigma'} . \tag{235}$$

All quantities involved in the contraction are labeled with spin indices. The CABS index is not present explicitly here (however, the CABS contribution can be implicitly included in the V intermediate, see Subsection 4.4 for the discussion about V) therefore only three spin cases ($\alpha\alpha$, $\beta\beta$ and $\alpha\beta$) have to be computed explicitly. The result of these contractions is then accumulated on the $^{\sigma\sigma'}\Omega^{BF}_{aibj}$ intermediate that refers to the proper spin case. The final contributions to the T_2 residual [Eqs. (184), (185) and (186)] are obtained from the $^{\sigma\sigma'}\Omega^{BF,\text{R12}}_{aibj}$ intermediate by contracting it with proper AO-MO transformation matrices. This is done for the entire BF contribution (conventional and explicitly correlated) and the implementation is part of the conventional code [48].

The $\Omega^{E',\text{R12}}_{xiyj}$ contribution to the $T_{2'}$ residual [Eq. (200)] is obtained by explicit contraction, the UHF version of this part has the following form

$$^{\sigma\sigma'}\Omega^{E',\text{R12}}_{xiyj} = \sum_{mn} B^{m\sigma,n\sigma'(ij)}_{x\sigma,y\sigma'} c^{m\sigma,n\sigma'}_{i\sigma,j\sigma'} . \tag{236}$$

The $^{\sigma\sigma'}\Omega^{B,\text{R12}}_{xiyj}$ and $^{\sigma\sigma'}\Omega^{F,\text{R12}}_{xiyj}$ contributions [Eqs. (197),(201)] are calculated by contracting the $M^{\mu,\nu}_{i,j}$ density with the V intermediate. The UHF analogue of this contraction can be written as

$$\sum_{\mu\nu} M^{\mu\sigma,\nu\sigma'}_{i\sigma,j\sigma'} V^{\mu\sigma,\nu\sigma'}_{x\sigma,y\sigma'} = \sum_{ab} t^{a\sigma,b\sigma'}_{i\sigma,j\sigma'} V^{\tilde{a}\sigma,\tilde{b}\sigma'}_{x\sigma,y\sigma'} + V^{\tilde{i}\sigma,\tilde{j}\sigma'}_{x\sigma,y\sigma'} = {}^{\sigma\sigma'}\Omega^{B,\text{R12}}_{xiyj} + {}^{\sigma\sigma'}\Omega^{F,\text{R12}}_{xiyj} , \tag{237}$$

with the following definition of the UHF $M^{\mu,\nu}_{i,j}$ density

$$M^{\mu\sigma,\nu\sigma'}_{i\sigma,j\sigma'} = \Lambda^{h,\sigma}_{\mu i} \Lambda^{h,\sigma'}_{\nu j} + \sum_{ab} \Lambda^{h,\sigma}_{\mu a} \Lambda^{h,\sigma'}_{\nu b} t^{a\sigma,b\sigma'}_{i\sigma,j\sigma'} . \tag{238}$$

The explicitly correlated C, D and E^1 contributions to the conventional doubles are obtained in the following way

$$^{\sigma\sigma'}\Omega^{C+D+E1,\text{R12}}_{aibj} = \sum_{p''k} \sum_{\sigma''} t^{a\sigma,p''\sigma''}_{i\sigma,k\sigma''} D'^{\sigma\sigma'}_{p''k,bj} - (1-\delta_{\sigma\sigma'}) \sum_{p''k} t^{p''\sigma,b\sigma'}_{i\sigma,k\sigma'} C'^{\sigma\sigma'}_{p''k,aj} + \tag{239}$$
$$\sum_{c} t^{a\sigma,c\sigma'}_{i\sigma,j\sigma'} E^{\sigma',1'}_{bc} ,$$

with the following definition of the intermediates

$$D'^{\sigma\sigma'}_{p''k,bj} = D^{\sigma\sigma'}_{p''k,bj} + \tfrac{1}{2}\delta_{\sigma\sigma'}\delta_{jk}E^{\sigma,1}_{bp''}, \tag{240}$$

$$C'^{\sigma\sigma'}_{p''k,bj} = C^{\sigma\sigma'}_{p''k,bj} - \tfrac{1}{2}\delta_{\sigma\sigma'}\delta_{jk}E^{\sigma,1}_{bp''}, \tag{241}$$

$$D^{\sigma\sigma'}_{p''k,bj} = \tilde{g}^{p''\sigma,j\sigma'}_{k\sigma,b\sigma'} - \delta_{\sigma\sigma'}\tilde{g}^{j\sigma,p''\sigma'}_{k\sigma,b\sigma'} + \frac{1}{2}\sum_{ld}\sum_{\sigma''} t^{b\sigma',d\sigma''}_{j\sigma',l\sigma''}\left(g^{p''\sigma,d\sigma''}_{k\sigma,l\sigma'} - \delta_{\sigma\sigma''}g^{p''\sigma,d\sigma''}_{l\sigma,k\sigma'}\right), \tag{242}$$

$$C^{\sigma\sigma'}_{p''k,bj} = \tilde{g}^{j\sigma,p''\sigma'}_{k\sigma,b\sigma'} - \frac{1}{2}\sum_{ld} t^{b\sigma,d\sigma'}_{l\sigma,j\sigma'}g^{q''\sigma,d\sigma'}_{l\sigma,k\sigma'}, \tag{243}$$

$$E^{\sigma,1'}_{bc} = -\sum_{lmp''}\sum_{\sigma'} t^{p''\sigma,b\sigma'}_{l\sigma,m\sigma'}\left(g^{p''\sigma,c\sigma'}_{l\sigma,m\sigma'} - \delta_{\sigma\sigma'}g^{c\sigma',p''\sigma}_{l\sigma,m\sigma'}\right), \tag{244}$$

$$E^{\sigma,1}_{bp''} = -F^{\sigma,val}_{\tilde{b}p''} + \sum_{k}\left(C^{\sigma\bar{\sigma}}_{p''k,bk} - D^{\sigma\sigma}_{p''k,bk}\right), \tag{245}$$

$$F^{\sigma,val}_{\tilde{b}p''} = \sum_{k} t^{b\sigma}_{k\sigma}F^{\sigma,val}_{kp''}. \tag{246}$$

The C intermediate is needed only for the opposite spin cases and the spin label with bar $\bar{\sigma}$ denotes the spin that is opposite to the spin σ. Similar to the RHF case, the traces of C and D are modified with the E^1 intermediates, yielding the primed intermediates. This is needed to account for for the Ω^{E1}_{aibj} term [Eq. (174)].
The UHF C, D and E^1 contributions to the F12 residual are obtained in the following way

$$^{\sigma\sigma'}\Omega^{C+D+E1,R12}_{xiyj} = \sum_{p''b}\sum_{\sigma''} f^{p''\sigma,b\sigma''}_{x\sigma,y\sigma''}\left(\sum_{cm} t^{b\sigma,c\sigma'}_{j\sigma,m\sigma'}D'^{\sigma\sigma'}_{cm,p''i} - (1-\delta_{\sigma\sigma'})\sum_{cm} t^{c\sigma,b\sigma'}_{j\sigma,m\sigma'}C'^{\sigma\sigma'}_{cm,p''i}\right), \tag{247}$$

with the intermediates defined as

$$D'^{\sigma\sigma'}_{cm,p''i} = D^{\sigma\sigma'}_{cm,p''i} + \tfrac{1}{2}\delta_{\sigma\sigma'}\delta_{ik}E^{\sigma,1}_{p''c}, \tag{248}$$

$$C'^{\sigma\sigma'}_{cm,p''i} = C^{\sigma\sigma'}_{cm,p''i} - \tfrac{1}{2}\delta_{\sigma\sigma'}\delta_{ik}E^{\sigma,1}_{p''c}, \tag{249}$$

$$D^{\sigma\sigma'}_{cm,p''i} = \tilde{g}^{c\sigma,i\sigma'}_{m\sigma,p''\sigma'} - \delta_{\sigma\sigma'}\tilde{g}^{i\sigma,c\sigma'}_{m\sigma,p''\sigma'}, \tag{250}$$

$$C^{\sigma\sigma'}_{cm,p''i} = \tilde{g}^{i\sigma',c\sigma}_{m\sigma',p''\sigma}, \tag{251}$$

$$E^{\sigma,1}_{p''c} = -F^{\sigma,val}_{p''c} + \sum_{k}\left(C^{\sigma\bar{\sigma}}_{p''k,ck} - D^{\sigma\sigma}_{p''k,ck}\right). \tag{252}$$

The discussion about the T_2 and $T_{2'}$ residuals requires the inclusion of the terms that involve the coupling matrices $C^{a,b}_{x,y}$. These terms, in the RHF formalism, were discussed in Subsection 4.8.1. The UHF analogue of Eqs. (227) and (228) can be written in the following form

$$^{\sigma\sigma'}\Omega^{E0}_{aibj} = \sum_{xy} C^{a\sigma,b\sigma'}_{x\sigma,y\sigma'}c^{x\sigma,y\sigma'}_{i\sigma,j\sigma'}, \tag{253}$$

$$^{\sigma\sigma'}\Omega^{E0}_{xiyj} = \sum_{ab} C^{a\sigma,b\sigma'}_{x\sigma,y\sigma'}t^{a\sigma,b\sigma'}_{i\sigma,j\sigma'}, \tag{254}$$

with the following UHF coupling matrix

$$C^{a\sigma,b\sigma'}_{x\sigma,y\sigma'} = \sum_{q''} f^{a\sigma,q''\sigma'\sigma'}_{x\sigma,y\sigma'}F_{q''b} + \sum_{p''} f^{p''\sigma,b\sigma'\sigma}_{x\sigma,y\sigma'}F_{ap''}. \tag{255}$$

In analogy to the RHF reference wave function, these terms are calculated in the indirect way by modifying the E^1 intermediate

$$E^{\sigma,1}_{ap''} \longleftarrow F^{\sigma,full}_{ap''}, \tag{256}$$

$$E^{\sigma,1}_{p''c} \longleftarrow F^{\sigma,full}_{p''c}. \tag{257}$$

All contributions to the coupled-cluster residuals discussed here (Subsections 4.7, 4.7.1, 4.8, 4.8.1, 4.8.2) together with the energy expressions (Subsection 4.5) were implemented in the **ricc2** program [51], which is part of the TURBOMOLE package [58]. The current level of implementation allows for computing the CCSD(F12) energies with the RHF, UHF and ROHF reference wave functions within the C_1 point group symmetry (which means that the molecules with higher point group symmetry are treated in the C_1 fashion).

4.9 Alternative approximations

The explicitly correlated contributions to the T_1, T_2 residuals and the $T_{2'}$ residual itself refer to the CCSD(F12) model, first introduced by Fliegl *et al.* [41]. This model is an approximation to the full CCSD-F12 approach where none of the F12 terms are neglected. The performance of the CCSD(F12) approximation has been extensively checked and it was shown that the simplifications made within this model only slightly affect the CC energies [59]. The intrinsic feature of this model is the need of recomputing the T_1-dependent V intermediate (discussed in Subsection 4.4.1) in each CC iteration. This quite soon becomes the bottleneck of the whole calculation, especially for the open-shell species where the $\alpha\beta$ components of the V intermediate are very expensive (due to the lowered symmetry of the integrals involving a CABS index). The T_1-dependent V intermediate is needed only for the evaluation of the $T_{2'}$ residual, thus avoiding the recalculation of this quantity within each CC iteration could possibly save a significant fraction of the CPU time in comparison to the CCSD(F12) model. Such situation can be reached within the orbital-invariant fixed-amplitudes approximation proposed recently by Tew *et al.* [60]. Within this approximation the CCSD(F12) energy is obtained from the coupled-cluster Lagrangian

$$E = \langle \text{HF}|\hat{H}^S|\text{HF}\rangle + \bar{\mathbf{t}}_1\langle\mu_1|\hat{H}^S|\text{HF}\rangle + \bar{\mathbf{t}}_2\langle\mu_2|\hat{H}^S|\text{HF}\rangle + \bar{\mathbf{t}}_{2'}\langle\mu_{2'}|\hat{H}^S|\text{HF}\rangle, \tag{258}$$

where the $\bar{\mathbf{t}}_1$, $\bar{\mathbf{t}}_2$ and $\bar{\mathbf{t}}_{2'}$ are the Lagrange multipliers and the Hamiltonian is similarity-transformed with $\hat{S} = \hat{T}_1 + \hat{T}_2 + \hat{T}_{2'}$. Within the orbital-invariant fixed-amplitudes approximation the $c^{x,y}_{i,j}$ coefficients are not optimized. Instead, their values are fixed according to the coalescence conditions for the first order pair functions [24, 25]. It means that only the elements $c^{i,j}_{i,j}$ and $c^{j,i}_{i,j}$ are considered. Moreover, their values are constant during the CC procedure. All the quantities that depend on these coefficients, and which were recomputed each iteration in the CCSD(F12) model (*e.g.* the F12 amplitudes in Eq. 157), now can be computed and stored only once, after the first CC iteration. The whole $T_{2'}$ residual is neglected, but the F12 contributions to the remaining residuals are included in the ordinary manner. Since the $c^{x,y}_{i,j}$ coefficients are not optimized, the $T_{2'}$ residual does not vanish. Therefore the correction to the electronic energy has to be computed. This takes place at the end of the CC procedure, when the T_1 and T_2 amplitudes are converged. The correction to the energy is computed by contracting the $c^{x,y}_{i,j}$ coefficients with the $T_{2'}$ residual

$$E^{\text{Lag}} = \sum_{xiyj} c^{x,y}_{i,j}\Omega_{xiyj}, \tag{259}$$

which requires a single evaluation of the $T_{2'}$ vector function at the end of the calculation. The final CCSD(F12/fixed) energy is the sum of the ordinary contribution [Eq. (124)] and the Lagrangian "penalty" term [Eq. (259)].

Recently Adler *et al.* proposed another approximation to the CCSD(F12) model, called CCSD-F12b, where even more F12 terms are neglected [61, 62]. Besides the aspects that refer to the fixed-amplitudes approximation, several additional simplifications are assumed. The only F12 contributions to the conventional doubles residual are the $\Omega_{aibj}^{B,\mathrm{R12}}$ [Eq. (185)] and Ω_{aibj}^{E0} [Eq. (227)] terms. The V intermediate used for the evaluation of the $\Omega_{aibj}^{B,\mathrm{R12}}$ term is computed in an approximated way by neglecting all the CABS contributions. The $T_{2'}$ residual needed to obtain the Lagrangian term [Eq. (259)] contains the $\Omega_{xiyj}^{E',\mathrm{R12}}$ [Eq. (200)] and Ω_{xiyj}^{E0} [Eq. (228)] contributions, while other ones are neglected. Within the present work, both approximated schemes have been implemented. In the case of the CCSD-F12b model, the inclusion of the Ω_{aibj}^{E0} and Ω_{xiyj}^{E0} terms would require programming the explicit contractions [Eqs. (227), (228)]. They were not necessary in the implementation of the CCSD(F12) model, because these terms were accounted for through the E^1 intermediates [Eqs. (230), (231)]. For these reasons they have not been included, and the model implemented in TURBOMOLE is dubbed CCSD-F12b*.

5 Accurate coupled-cluster calculations of the reaction barrier heights of two $CH_3^\bullet$+CH_4 reactions

We have computed barrier heights of 71.8 ± 2.0 kJ mol^{-1} and 216.4 ± 2.0 kJ mol^{-1} for the reactions $CH_4 + CH_3^\bullet \rightleftarrows CH_3 \cdots H \cdots CH_3^\bullet \rightarrow CH_3^\bullet + CH_4$ and $CH_4 + CH_3^\bullet \rightleftarrows H \cdots C_2H_6^\bullet \rightarrow H^\bullet + C_2H_6$, respectively, using explicitly correlated coupled-cluster theory with singles and doubles combined with standard coupled-cluster theory with up to connected quadruple excitations. Transition-state theory has been used to compute the respective reaction rate constants in the temperature interval 250–1500 K. The computed rates for the reaction to ethane are orders of magnitude slower than those used in the mechanism of Norinaga and Deutschmann (Ind. Eng. Chem. Res. **46** (2007) 3547) for the modeling of the chemical vapor deposition of pyrolytic carbon.

5.1 Introduction

Recently, Norinaga and Deutschmann modeled the chemical kinetics of the pyrolysis of the hydrocarbons ethylene, acetylene, and propylene under conditions relevant to the chemical vapor deposition (CVD) of pyrolytic carbon [63]. For this modeling, they had developed a mechanism containing 227 species and 827 reactions. One of these was

$$CH_4 + CH_3^\bullet \rightarrow C_2H_6 + H^\bullet, \qquad (260)$$

which contributed to the mechanism with the reaction rate constant

$$k = AT^n \exp\left(\frac{-E_a}{RT}\right), \qquad (261)$$

with $A = 8.0 \times 10^{13}$ cm^3 mol^{-1} s^{-1}, $n = 0$, and $E_a = 167.37$ kJ mol^{-1}. The parameters were taken from the work of Tabayashi and Bauer [64]. Note that these parameters are fit parameters that were adjusted by requiring the best overall agreement between observed density-gradient profiles and those obtained from calculations based on a twelve-step decomposition mechanism for the pyrolysis of methane [64]. Reaction (260) was considered by Kassel [65] almost 75 years ago when discussing the role of methyl and methylene radicals in the decomposition of methane. The reaction was also discussed in 1959 by Skinner and Ruehrwein [66], who crudely estimated the rate constant as $k = 10^9 T \exp\{-E_a/(RT)\}$ cm^3 mol^{-1} s^{-1} with $E_a = 188.28$ kJ mol^{-1} (average of 40–50 kcal mol^{-1}). Based on this estimate, these authors concluded that the reaction can be neglected in a mechanism for the pyrolysis of methane [66]. Furthermore, more than 25 years ago, Back [67] derived an upper limit for the reaction rate constant of $k = 63$ cm^3 mol^{-1} s^{-1} at 802 K, about 16 times below the value of Tabayashi and Bauer for that same temperature. Hence, the role of this reaction in the mechanism of Norinaga and Deutschmann was considered to be suspicious, and we decided to reinvestigate reaction (1) by means of modern, high-level *ab initio* quantum chemical methods.

In the present work, we show that the production of ethane from the reaction of methane with methyl is grossly overestimated by the above rate constant. From transition-state theory (TST) and highly correlated *ab initio* calculations, we find rates that are between three (at 1500 K) and ten (at 300 K) orders of magnitudes slower than those used by Norinaga and Deutschmann [63]. This leads us to conclude that the role of this reaction in the mechanism must be reexamined. Furthermore, we show that the barrier height for the above reaction to C_2H_6 and $H^\bullet$ is much higher (about three times higher) than the barrier height for another reaction between the two reactants, the reaction in which a hydrogen is abstracted from methane by a methyl radical.

A second purpose of the present work is to assess the performance of the explicitly correlated coupled-cluster model CCSD(F12) that we have recently implemented in the TURBOMOLE program package [68, 69]. This model has the potential to yield electronic molecular energies at the level of coupled-cluster theory with single and double excitations (CCSD [37, 70]) at the limit of a complete one-particle basis set. In conjunction with corrections for higher excitations (connected triples and connected quadruples) it should be possible to compute the barrier height for the above reaction with an accuracy of about 1–2 kJ mol^{-1}, that is, with an error of about 0.5–1.0%.

The CCSD(F12) model was first introduced as CCSD(R12) approximation [41] to the full CCSD-R12 approach [40]. Here, we indicate by writing "R12" that in this initial work, explicitly correlated two-particle basis functions of the form

$$\chi_{ij}(1,2) = r_{12}\,\varphi_i(1)\varphi_j(2) \tag{262}$$

were used, where $r_{12} = |\mathbf{r}_1 - \mathbf{r}_2|$ is the distance between the electrons 1 and 2, and where φ_i and φ_j are two spin orbitals that are occupied in the Hartree–Fock reference determinant. More recently, the R12-type two-particle basis functions were replaced by functions of the type

$$\chi_{ij}(1,2) = f(r_{12})\,\varphi_i(1)\varphi_j(2) = \exp(-\gamma r_{12})\,\varphi_i(1)\varphi_j(2), \tag{263}$$

also called Slater-type geminals (STG) [45]. We will refer to the corresponding approach as CCSD(F12) model [31, 32].

The CCSD(F12) model as well as the full CCSD-F12 approach and other simplifications of it are currently being implemented in various quantum chemistry programs [59, 61, 62, 71, 72, 73, 74, 75, 76, 77, 78, 79], also in combination with connected triples and higher excitations. In particular Köhn and co-workers [72] have shown that the CCSD(F12) model is an excellent approximation to the full CCSD-F12 approach, and the CCSD(F12) model is the method of choice that we have implemented in the TURBOMOLE program. The present work reports on one of the first applications of CCSD(F12) theory with chemical relevance. In such a real-life application, CCSD(F12) calculations are combined with a series of other coupled-cluster calculations including geometry optimizations, calculations of harmonic vibrational frequencies, and coupled-cluster calculations with connected triples and quadruples. Within the whole set of calculations that must be performed, the CCSD(F12) calculations take only a fraction of the total computation time, and therefore, in an application as the one presented here, there appears to be no need to further simplify the CCSD(F12) model.

The present work is organized as follows: In Subsection 5.2, we present the details of the *ab initio* calculations of the reaction barriers (Subsection 5.2.1) and of the reaction rate constants by means of transition-state theory (Subsection 5.2.2). Accordingly, the results are presented in Subsection 5.3, both for the reaction barriers (Subsection 5.3.1) and the reaction rate constants (Subsection 5.3.2). Conclusions are collected in Subsection 5.4.

5.2 Computational Details

5.2.1 Reaction Barriers

The geometries of $CH_3^\bullet$, CH_4, C_2H_6, $H \cdots C_2H_6^\bullet$ and $CH_3 \cdots H \cdots CH_3^\bullet$ were optimized at the all-electron (ae) CCSD(T)/aug-cc-pwCVTZ level[80, 81, 82] using the ACES II program package [83]. For the open-shell systems, an unrestricted Hartree–Fock (UHF) reference determinant was used. The optimized equilibrium C–H distances in $CH_3^\bullet$, CH_4 and C_2H_6 are 107.78, 108.80 and 109.08 pm, respectively. In ethane, the optimized C–C bond amounts to 152.60 pm and the C–C–H angle to 111.20°. The equilibrium structures of the two transition states are depicted in Figure 5. Our equilibrium structures of $[H \cdots C_2H_6^\bullet]^\ddagger$

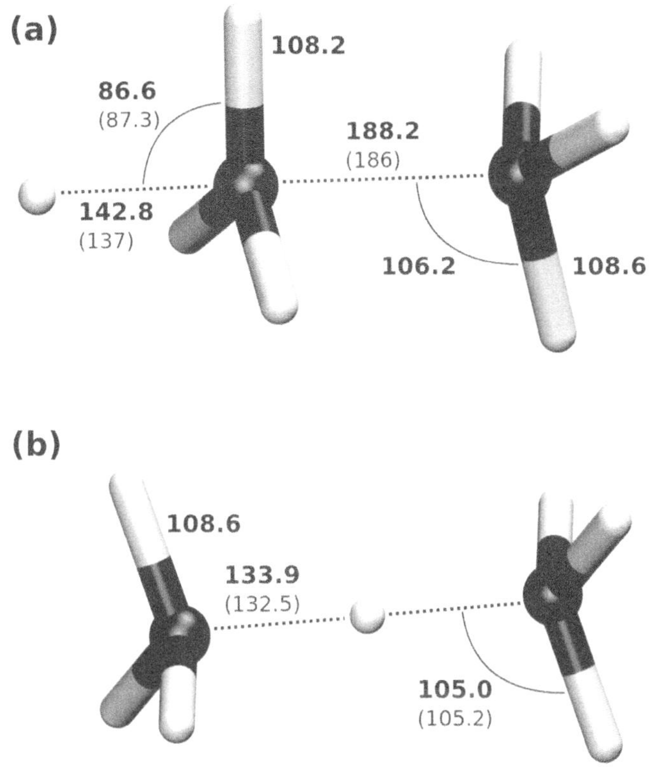

Figure 5: Equilibrium structures of the transition structures (a) $[H \cdots C_2H_6^\bullet]^\ddagger$ (C_{3v} symmetry) and (b) $[CH_3 \cdots H \cdots CH_3^\bullet]^\ddagger$ (D_{3d} symmetry) as obtained at the ae-CCSD(T)/aug-cc-pwCVTZ level. Geometry parameters from the work of Layfield *et al.* [84] and Remmert *et al.* [85] are shown between parentheses. Bond lengths are given in pm and angles in degrees.

and $[CH_3 \cdots H \cdots CH_3^\bullet]^\ddagger$ compare well with those obtained by Layfield *et al.* [84] and Remmert and co-workers [85], respectively.

Single-point energy calculations were performed at the ae-CCSD(T)/aug-cc-pwCVQZ′ level, where the prime indicates that the aug-cc-pwCVQZ basis was used for C but only the aug-cc-pVTZ basis for H. For comparison, frozen-core (fc) calculations were performed to quantify core-valence (CV) correlation effects. The 1*s* orbitals of C were kept frozen in the fc calculations.

Further single-point energy calculations were performed at the fc-CCSDT/aug-cc-pVDZ and fc-CCSDT(Q)/cc-pVDZ levels [38, 39, 86, 87, 88] using the MRCC program package [89, 90].
Explicitly correlated coupled-cluster calculations were performed at the fc-CCSD(F12)/cc-pVQZ-F12 level using the TURBOMOLE program package [69]. The exponent of the Slater-type geminal was $\gamma = 1.1\ a_0^{-1}$, which is the optimal exponent for the cc-pVQZ-F12 basis set [91]. We employed the complementary auxiliary basis set approach (CABS) of Valeev [20] using the auxiliary basis set of Yousaf and Peterson [92] optimized for the cc-pVQZ-F12 atomic orbital basis set. For the two-electron integrals, computed employing the density-fitting technique [21], we used the aug-cc-pwCV5Z MP2 fitting basis of of Hättig [93]. To represent the Fock operator within the CCSD(F12) calculations, we used the aug-cc-pV5Z JK fitting basis of Weigend [94]. The orbital-invariant version of CCSD(F12) theory was used [31].
The cc-pVDZ, aug-cc-pVXZ (X = D, T, Q) and aug-cc-pwCVXZ (X = T, Q) basis sets [95, 96, 97] were obtained from the Basis Set Library of the Environmental Molecular Sciences Laboratory [98, 99]. The cc-pVQZ-F12 sets with corresponding CABS were downloaded from the web site of Peterson [100].
Corrections for scalar-relativistic effects (one-electron Darwin and mass-velocity terms, MVD) were calculated at the ae-CCSD(T)/aug-cc-pwCVTZ level [101, 102]. For the C atom, the spin–orbit correction (SO) to the total electronic energy amounts to $\Delta E_{\mathrm{SO}} = -0.35399\ \mathrm{kJ\,mol^{-1}}$ [103].

5.2.2 Reaction Rates

Simple transition state theory is used for the calculation of the reaction rate constants. The well-known expression for the reaction rate constant of a bimolecular reaction $\mathrm{X} + \mathrm{Y} \rightleftarrows \mathrm{XY}^\ddagger$ is [104]

$$k = \kappa \frac{k_B T}{h} \frac{(q^\ddagger_{\mathrm{XY}}/V)}{(q_{\mathrm{X}}/V)(q_{\mathrm{Y}}/V)} \exp\left(\frac{-\Delta E_{B,0}}{RT}\right), \tag{264}$$

where $q^\ddagger_{\mathrm{XY}}$, q_{X} and q_{Y}, are the dimensionless partition functions (including translational, vibrational and rotational contributions) of the transition state and the reactants, respectively (using the harmonic oscillator, rigid rotor approximation and correcting for internal hindered rotations). The degeneracies of the doublet states are not considered since they cancel out in Eq. (264). R is the gas constant, k_B the Boltzmann constant, h the Planck constant and V the volume that cancels the volume in the translational partition function,

$$q_{\mathrm{trans}} = \left(\frac{2\pi m k_B T}{h^2}\right)^{3/2} V, \tag{265}$$

where m is the mass of the molecule or radical. $\Delta E_{B,0}$ is the electronic barrier height $\Delta E_{B,e}$ plus the zero-point vibrational energy (ZPVE). The ZPVE is computed at the ae-CCSD(T)/aug-cc-pwCVTZ level. κ is the transmission coefficient accounting for tunneling effects, computed from the Wigner formula [105]

$$\kappa = 1 - \frac{1}{24}\left(\frac{h\nu}{k_B T}\right)^2 \left(1 + \frac{RT}{\Delta E_{B,0}}\right). \tag{266}$$

Only the imaginary frequency ν, associated with the reaction coordinate, and the reaction barrier $\Delta E_{B,0}$ are required to calculate κ. At the ae-CCSD(T)/aug-cc-pwCVTZ level, the imaginary frequencies are $1540.4i$ and $1855.9i$ cm^{-1} for $H\cdots C_2H_6^\bullet$ and $CH_3\cdots H\cdots CH_3^\bullet$, respectively.

Corrections accounting for hindered rotations are included for rotations about the C–C bond and about the reaction coordinate C$\cdots$H$\cdots$C. To obtain these corrections, we computed the potential energy of the hindered internal rotor,

$$V(\phi) = \frac{V_s}{2}\left[1 - \cos(3\phi)\right], \quad 0 \leq \phi \leq 2\pi, \tag{267}$$

at the level of ae-CCSD(T)/aug-cc-pwCVTZ theory, optimizing all geometry parameters except ϕ. At this level, the barrier heights V_s are 11.74, 6.56 and 0.22 kJ mol^{-1} for C_2H_6, $H\cdots C_2H_6^\bullet$ and $CH_3\cdots H\cdots CH_3^\bullet$, respectively. Each reduced moment of inertia was computed from the distances (in the equilibrium geometry) of the rotating H atoms from the axis of rotation. These distances amount to 101.70, 104.28/108.01 and 104.84 pm for C_2H_6, $[H\cdots C_2H_6^\bullet]^\ddagger$ and $[CH_3\cdots H\cdots CH_3^\bullet]^\ddagger$, respectively (Table 2). The eigenvalues of the Schrödinger

Table 2: Rotational barriers (V_s) and constants (B).

System	V_s/kJ mol^{-1}	R^a_{XH}/pm	B/kJ mol^{-1}	Θ^b_{tors}/K
C_2H_6	11.74	101.70	0.129	447.7
$[CH_3\cdots H\cdots CH_3^\bullet]^\ddagger$	0.22	104.84	0.121	61.2
$[H\cdots C_2H_6^\bullet]^\ddagger$	6.56	104.28	0.118	321.0
		108.01		

[a]Distance of the H atom from the axis of rotation.
[b]Characteristic vibrational temperature of the torsional vibration.

equation with the potential Eq. (267) were obtained by diagonalizing a tridigonal matrix of dimension 501, following recent work by Strekalov [106]. In Eq. (264), the vibrational partition functions are computed with respect to the corresponding lowest vibrational level ($v = 0$) as zero of energy. This yields

$$q_{vib} = \prod_{k=0}^{n}\left[1 - \exp(-\Theta_k/T)\right]^{-1}, \tag{268}$$

where the product runs over all real harmonic frequencies with characteristic vibrational temperatures Θ_k. This vibrational partition function is corrected for hindered internal rotation by multiplying by the factor

$$q_{hr}/q_{tors} = q_{hr}\left[1 - \exp(-\Theta_{tors}/T)\right], \tag{269}$$

where q_{hr} is the partition function of the hindered internal rotor and q_{tors} the partition function of the harmonic torsional vibration, both evaluated with respect to their own lowest level as zero of energy. The difference between the zero-point energies of the one-dimensional harmonic oscillator and the hindered internal rotor is taken into account when calculating the electronic energy (column "HR" in Table 3).

5.3 Results

5.3.1 Reaction Barriers

For the five species involved in this study, the computed electronic atomization energies (AE, the dissociation energy $\sum D_0$ for the dissociation into isolated atoms) are given in Table 3 and compared with literature values [107, 108] where possible. The latter originate

Table 3: Electronic atomization energies[a] / kJ mol^{-1}.

System	CCSD	(T)	(Q)	CV	ZPVE	HR	MVD	SO	Total	ATcT
$CH_3^{\bullet}$	1275.1	7.6	0.5	4.5	−78.3	...	−0.7	−0.4	1208.4	1209.7[b]
CH_4	1741.0	11.9	0.5	5.3	−117.9	...	−0.8	−0.4	1639.6	1642.2[c]
C_2H_6	2945.8	26.2	0.9	10.3	−196.7	0.1	−1.7	−0.7	2784.3	2787.2[c]
$[CH_3^{\bullet}\cdot H\cdots CH_3^{\bullet}]^{\ddagger}$	2933.2	27.7	1.9	9.5	−194.0	0.2	−1.6	−0.7	2776.2	
$[H\cdots C_2H_6^{\bullet}]^{\ddagger}$	2787.2	33.4	2.6	8.9	−198.4	0.1	−1.6	−0.7	2631.6	

[a]The CCSD energies were obtained at the fc-CCSD(F12)/cc-pVQZ-F12 level. The corrections for connected triple excitations (T) were obtained at the fc-CCSD(T)/aug-cc-pwCVQZ′ level. The correction (Q) for connected quadruples contains the difference CCSDT–CCSD(T) calculated at the fc-CCSDT/aug-cc-pVDZ level and the (Q) term obtained at the fc-CCSDT(Q)/cc-pVDZ level. The correction for core–valence correlation (CV) was obtained at the ae-CCSD(T)/aug-cc-pwCVQZ′ level. The harmonic zero-point vibrational energy (ZPVE) correction is supplemented by a zero-point-energy correction for hindered rotation (HR), and both scalar relativistic (MVD) and spin–orbit (SO) effects are taken into account.
[b]Taken from the work of Aguilera-Iparraguirre and co-workers [107].
[c]Taken from the work of Klopper and co-workers [108].

from the Active Thermochemical Tables (ATcT) of Ruscic [109, 110, 111, 112].
The computed AEs are found 1.3, 2.6, and 2.9 kJ mol^{-1} below the ATcT reference values for $CH_3^{\bullet}$, CH_4, and C_2H_6, respectively. One reason for these discrepancies is that in the present work, we have only included (except for the hindered-rotor treatment) the *harmonic* ZPVE. For example, the anharmonic correction to the ZPVE of $CH_3^{\bullet}$ amounts to 0.9 kJ mol^{-1} at the ae-CCSD(T)/cc-pCVTZ′ level [107]. Taking this anharmonic correction into account would have produced a theoretical AE for $CH_3^{\bullet}$ of 1209.3 kJ mol^{-1}, within 0.4 kJ mol^{-1} of the experimental value. The total ZPVE contribution would have been −77.4 kJ mol^{-1}, in good agreement with the value (−77.6 kJ mol^{-1}) of Schwenke [113]. For methane, Schwenke's value [114] amounts to −116.1 kJ mol^{-1}, whereas our harmonic value is −117.9 kJ mol^{-1}. Taking Schwenke's value in place of ours would reduce the error in the calculated AE of methane from 2.6 to 0.8 kJ mol^{-1}. For ethane, an accurate (anharmonic) ZPVE contribution of −194.1 kJ mol^{-1} is available from benchmark calculations performed by Karton and co-workers [115]. This contribution is 2.6 kJ mol^{-1} smaller in magnitude than our harmonic value, which makes up for almost all of the error of 2.9 kJ mol^{-1}. Furthermore, our fc-CCSD(T) value of 2972.5 kJ mol^{-1} compares well with the value of 2973.7 kJ mol^{-1} obtained at the W4 level [116] by these authors.
We expect that the errors due to neglecting the anharmonic corrections to the ZPVE will

largely cancel when we compute the relative quantities of interest such as reaction energies and barrier heights. The calculated energy of reaction for the endothermic reaction

$$CH_4 + CH_3^\bullet \rightarrow C_2H_6 + H^\bullet, \tag{270}$$

for instance, amounts to 63.7 kJ mol^{-1}, only 1.0 kJ mol^{-1} below the ATcT value of 64.7 kJ mol^{-1}. The anharmonic corrections to the transition states with their partially broken bonds may be somewhat larger than for the molecules $CH_3^\bullet$, CH_4, and C_2H_6 but are difficult to quantify. The lowest-energy vibrational modes not accounted for in the hindered-rotor treatments have wavenumbers 317 cm^{-1} ($CH_3 \cdots H \cdots CH_3^\bullet$) and 640 cm^{-1} ($H \cdots C_2H_6^\bullet$). Scaling all of the harmonic vibrational frequencies by a factor of 0.987, which yields an anharmonic correction to the AEs of $CH_3^\bullet$, CH_4, and C_2H_6 of the correct order of magnitude, would affect the reaction barrier heights only by ca. 0.03 kJ mol^{-1} and the $\log(k)$ values in Table 4 only by about 0.02.

Similar error cancellations occur for other (small) ignored effects and remaining errors. For example, the diagonal Born-Oppenheimer correction (DBOC) can crudely be estimated to contribute *ca.* 0.1 kJ mol^{-1} per CH bond to the AEs of the systems studied here [108], and can thus safely be neglected when computing reaction energies and barrier heights. Another example is the basis set error in the (T) corrections, which can be estimated from an extrapolation from aug-cc-pwCVQZ′ and aug-cc-pwCVTZ results using the X^{-3} formula of Helgaker and co-workers [117]. This increases the (T) contributions to the AEs by 0.2 kJ mol^{-1} for $CH_3^\bullet$ and CH_4, and by 0.5 kJ mol^{-1} for C_2H_6, $H \cdots C_2H_6^\bullet$, and $CH_3 \cdots H \cdots CH_3^\bullet$. (Note that the (T) contribution for ethane computed by Karton *et al.* [115] is also *ca.* 0.5 kJ mol^{-1} larger than ours.) Thus, the effect of the (T) basis set error on the relevant relative quantities is only of the order of 0.1 kJ mol^{-1} and thus negligible. The cancellation of small terms is even more striking for the Hartree–Fock corrections that are contained in our CCSD(F12) values. These corrections are computed as second-order perturbation theory corrections from single excitations into the complementary auxiliary basis set (CABS) that is used in the CCSD(F12) model. The corresponding contributions to the AEs amount to 0.14, 0.18, 0.29, 0.31, and 0.31 kJ mol^{-1} for $CH_3^\bullet$, CH_4, C_2H_6, $[H \cdots C_2H_6^\bullet]^\ddagger$, and $[CH_3 \cdots H \cdots CH_3^\bullet]^\ddagger$, respectively, with virtually no net effect on the reaction energies and barrier heights. These Hartree–Fock corrections could just as well have been omitted.

Also the other terms are either so accurate or so small that we can rely on a cancellation of the remaining errors in these terms. Concerning the CV correction, our values for $CH_3^\bullet$, CH_4, and C_2H_6 (4.5, 5.3, and 10.3 kJ mol^{-1}) agree well with the corresponding values from W4 theory (4.56, 5.31, and 10.2 kJ mol^{-1}) [115, 116]. The sum of our CCSD, (T), and CV terms gives an atomization energy of 1287.4 kJ mol^{-1} for $CH_3^\bullet$ (after adding 0.2 kJ mol^{-1} from extrapolating the (T) correction). This compares favorably with the value of 1288.0 kJ mol^{-1} obtained at the HEAT-456 level by Harding *et al.* [118], which contains inner-shell correlation. Our MVD contributions agree to within 0.05 kJ mol^{-1} with those of the W4 and HEAT protocols [115, 116, 118]. Correlation effects beyond the CCSD(T) level, however, seem somewhat larger (*e.g.*, by about a factor of two for ethane [115]) in our calculations than in the W4 and HEAT-456(Q) protocols, but the corresponding contributions to the reaction energies and barrier heights are very small.

In view of all of the above considerations, we expect that our computed reaction barrier heights are accurate to within 2 kJ mol^{-1}. Our final values thus are $\Delta E_{B,0} = 71.8 \pm 2.0$ kJ mol^{-1} and $\Delta E_{B,0} = 216.4 \pm 2.0$ kJ mol^{-1} for the reactions $CH_4 + CH_3^\bullet \rightarrow CH_3^\bullet + CH_4$ and $CH_4 + CH_3^\bullet \rightarrow H^\bullet + C_2H_6$, respectively. Concerning the latter reaction, we obtain $\Delta E_{B,0} = 152.7 \pm 2.0$ kJ mol^{-1} for the backward reaction, remarkably close but slightly below the CCSD(T)/aug-cc-pVQZ//MP2/aug-cc-pVDZ value of 155.0 kJ mol^{-1} computed (presumably in the frozen-core approximation) by Layfield and co-workers [84]. Also our computed barrier height for the reaction $CH_4 + CH_3^\bullet \rightarrow CH_3^\bullet + CH_4$ (71.8 ± 2.0 kJ mol^{-1}) is slightly below those obtained by Remmert *et al.* [85] (74.1 kJ mol^{-1} at the CCSD(T)/cc-pVTZ//MP2/cc-pVTZ level) and Kungwan and Truong[119] (75.7 kJ mol^{-1} at the CCSD(T)/cc-pVTZ//BH&HLYP/cc-pVDZ level).

5.3.2 Reaction Rates

Computed reaction rate constants (or rather the logarithm thereof) are shown in Table 4 for the reactions $CH_4 + CH_3^\bullet \rightarrow CH_3^\bullet + CH_4$ (reaction **1**) and $CH_4 + CH_3^\bullet \rightarrow H^\bullet + C_2H_6$ (reaction **2**). They are also depicted in Figure 6, which not only shows the calculated

Table 4: Logarithm of the reaction rate constant, $\log(k/[\mathrm{cm^3\,mol^{-1}\,s^{-1}}])$, as a function of temperature, for the reactions $CH_4 + CH_3^\bullet \rightarrow CH_3^\bullet + CH_4$ (reaction **1**) and $CH_4 + CH_3^\bullet \rightarrow H^\bullet + C_2H_6$ (reaction **2**). The transmission coefficient κ accounting for tunneling effects is also given.

T / K	Reaction **1**		Reaction **2**		
	This work	κ	This work	κ	Used in Ref. [63]
300	0.20	4.42	−25.38	3.30	−15.24
400	3.13	2.94	−16.18	2.30	−7.95
500	4.92	2.26	−10.63	1.83	−3.58
600	6.14	1.88	−6.89	1.58	−0.67
700	7.05	1.66	−4.20	1.43	1.41
800	7.75	1.51	−2.15	1.33	2.98
900	8.31.	1.41	−0.54	1.26	4.19
1000	8.78	1.33	0.76	1.21	5.16
1100	9.17	1.28	1.84	1.18	5.96
1200	9.51	1.23	2.75	1.15	6.62
1300	9.81	1.20	3.53	1.13	7.18
1400	0.07	1.18	4.21	1.11	7.66
1500	0.30	1.15	4.80	1.10	8.07

rate constants at 300, 350, ..., 1500 K, but also the curves that are obtained by fitting expressions of the type of Eq. (261) to the calculated points in the range 600 − 1500 K. In Figure 6, the solid line corresponds to the reaction rate constants used in Ref. [63], that is, using Eq. (261) with the parameters of Tabayashi and Bauer [64]. Our rate constants are below the upper limit of Back [67] at 802 K as well as below the rate constant expression of Skinner and Ruehrwein [66].
The parameters A, n, and E_a, obtained by fitting computed rate constants to the expression Eq. (261), are given in Table 5, not only for reactions **1** and **2**, but also for the backward reaction of **2**.

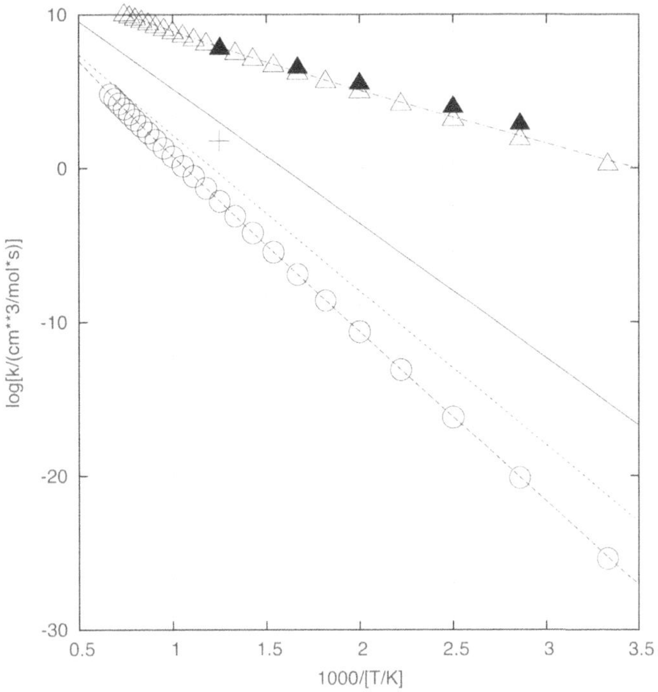

Figure 6: Reaction rate constants (k in $cm^3 mol^{-1} s^{-1}$) as a function of $1/T$ for the reactions $CH_4 + CH_3^\bullet \rightarrow CH_3^\bullet + CH_4$ (△) and $CH_4 + CH_3^\bullet \rightarrow H^\bullet + C_2H_6$ (○). For comparison, the rate constant used in Ref. [63] for the latter reaction are shown as solid line. The upper limit of Back [67] at 802 K is marked as +. The rate constants of Skinner and Ruehrwein [66] are indicated as dotted line. The evaluated experimental kinetic data of Kerr and Parsonage [120] for the reaction $CH_4 + CH_3^\bullet \rightarrow CH_3^\bullet + CH_4$ are shown as ▲.

Table 5: Fitting parameters obtained by adjusting the rate constant equation $k = AT^n \exp\{-E_a/(RT)\}$ to the calculated values in the range 600 – 1500 K.

Reaction	$A/cm^3 mol^{-1} s^{-1}$	n	$E_a/kJ mol^{-1}$
$CH_4 + CH_3^\bullet \rightarrow CH_3^\bullet + CH_4$	1.5×10^2	3.2	55
$CH_4 + CH_3^\bullet \rightarrow H^\bullet + C_2H_6$	4.3×10^1	3.2	200
$H^\bullet + C_2H_6 \rightarrow CH_4 + CH_3^\bullet$	1.2×10^8	1.7	145

The reaction rate constants for reaction **2** as computed in the present work are 3–10 orders of magnitude smaller than those of Tabayashi and Bauer [64]. We also note that the calculated points are very well represented by the fits.

We can compare our computed reaction rate constants for reaction **1** ($CH_4 + CH_3^\bullet \rightarrow CH_3^\bullet + CH_4$) with the accurate results obtained very recently for this reaction by Remmert *et al.* by means of a reduced dimensionality quantum dynamics study [85]. These authors compare their results with evaluated kinetic data of Kerr and Parsonage [120] as well as with those obtained by Kungwan and Truong [119] from canonical variational transition-state theory (CVT) including corrections for hindered rotation and tunneling. The latter correction was obtained using the multidimensional semiclassical small-curvature tunneling method (SCT) of Truhlar and co-workers [121]. Figure 6 shows that our computed ate constants for reaction **1** are slightly below those of Kerr and Parsonage [120]. This is consistent with the TST results reported by Remmert *et al.* [85] and Kungwan and Truong [119], which also are slightly below the evaluated kinetic data of Kerr and Parsonage [120]. Remmert *et al.* [85] note that tunneling plays an important role in the low temperature region. The tunneling correction obtained from the Wigner formula, however, is significantly smaller than the one obtained from the SCT method (see Table 4 and Ref. [119]). At 600, 800, and 1000 K, for example, the SCT tunneling corrections are 1.7, 1.3, and 1.1 times larger than the Wigner values. Nevertheless, it was more the purpose of the present work to compute accurate reaction barriers heights than to compute accurate reaction rate constants. In this respect and in particular in view of the uncertainty in our tunneling corrections, the agreement between our computed rate constants and the experimental values for reaction **1** is satisfactory.

5.4 Conclusions

We have determined barrier heights of 71.8 ± 2.0 kJ mol^{-1} and 216.4 ± 2.0 kJ mol^{-1} for the reactions $CH_4 + CH_3^\bullet \rightarrow CH_3^\bullet + CH_4$ and $CH_4 + CH_3^\bullet \rightarrow H^\bullet + C_2H_6$, respectively, from benchmark *ab initio* calculations. Using these barrier heights in conjunction with simple transition-state theory yields rate constants for the latter reaction that are orders of magnitude smaller than the ones used in the mechanism of Ref. [63], for the modeling of the pyrolysis of ethylene, acetylene, and propylene. Therefore, we suggest that it is necessary to reinvestigate the role of this reaction in that mechanism. Furthermore, we suggest to use Eq. (261) with the parameters $A = 43$ cm^3 mol^{-1} s^{-1}, $n = 3.2$, and $E_a = 200$ kJ mol^{-1} for the reaction $CH_4 + CH_3^\bullet \rightarrow H^\bullet + C_2H_6$.

6 Calculating atomization energies accurately without falling back on extrapolations or empirical corrections

The atomization energies of the 105 molecules in the test set of Bakowies [J. Chem. Phys. **127**, 084105 (2007)] have been computed with an estimated standard deviation (from the values compiled in the Active Thermochemical Tables) of ±0.13 kJ/mol per valence electron in the molecule. The equilibrium geometries and harmonic frequencies were obtained at the CCSD(T) level of theory with the cc-pCVTZ basis set. In order to obtain the final atomization energies accurately, the following contributions have been taken into account: the correction to the basis set truncation error (obtained at the CCSD(F12) level), the correction for anharmonicity effects and zero-point vibrational energy, relativistic correction, the correction accounting for the improvement between the CCSD(T) and full CCSDT coupled-cluster models and the correction for the perturbative treatment of the connected quadruple excitations (Q). The good agreement with the experimental values (standard deviation of ±0.13 kJ/mol per valence electron) was obtained here without falling back on empirical corrections.

6.1 Introduction

The purpose of the present work is to test the efficiency and performance of the implementation of the explicitly correlated coupled-cluster singles-and-doubles, CCSD(F12), model in the TURBOMOLE package. This model, initially proposed by Fliegl *et al.* [41], is the approximation to the full CCSD-F12 approach originally given by Noga *et al.* [40]. It has been demonstrated that this method is able to recover quintuple-ζ quality correlation energies with the orbital basis set of the size of triple-ζ [31].
The accurate *ab initio* estimation of the thermodynamic quantities requires taking into account various effects. Besides the contributions from the levels of theory that refer to the hierarchy of coupled-cluster models, also the effects accounting for relativistic, anharmonic and other phenomena should be taken into account. Clearly, the current level of a development of the theoretical molecular sciences does not allow obtaining the final, accurate result within one single calculation. Instead, one has to combine a couple (sometimes dozens) of calculations to determine the final value in an additivity scheme. Various such additivity schemes and model chemistries already exist in the literature. For instance the Gaussian-n ($n = 2, 3, 4$) theories of Curtis *et al.* [122, 123, 124], the correlation-consistent composite approach (ccCA) of DeYonker *et al.* [125, 126], the complete basis set (CBS) method of Petersson and co-workers [127, 128], the focal point analysis (FPA) proposed by Allen *et al.* [129, 130], the multicoefficient correlation method (MCCMs) of Fast *et al.* [131, 132, 133], the HEAT protocol by Tajti *et al.* [134], Bomble *et al.* [135] and Harding *et al.* [118], and the Weizmann-n ($n = 1 - 4$) theories of Martin and co-workers [136, 137, 116, 138]. The general idea of all of these schemes is similar, they combine calculations at various levels of theory and the final quantity is constructed as the sum of the incremental contributions. Sometimes scaling factors or extrapolation techniques are used.
The present application of the CCSD(F12) model is an extension of the previous work of Klopper *et al.* [108], where the set of molecules, initially proposed by Bakowies [139], was used for testing the performance of the implementation of the MP2-F12 method in

the TURBOMOLE program [49, 50]. To get agreement with the experimental atomization energies, the authors in Ref. [108] applied an empirical "interference" factor for the correction to the basis set truncation error. Here we report the results of similar accuracy obtained without such empirical factors.

6.2 Methodology

All contributions to the atomization energies, except the anharmonic correction, were obtained at the ae-CCSD(T)/cc-pCVTZ equilibrium geometries. The anharmonic correction, taken from Ref. [108], was calculated at the fc-MP2/cc-pVDZ optimal geometry. Our proposed incremental scheme begins with the coupled-cluster singles-and-doubles (CCSD) contribution in the correlation-consistent core-valence quadruple-zeta basis set (cc-pCVQZ) [140], correlating only the valence electrons (frozen-core approximation, fc-CCSD). The atomization energies obtained at this level (fc-CCSD/cc-pCVQZ) are reported under "CCSD" in Table 7. All molecules considered within the present study are closed-shell systems, and restricted Hartree–Fock (RHF) wave functions were used as the references. The electronic energies referring to the open-shell atoms (H, C, N, O, F), needed for the evaluation of the atomization energies, were obtained within spin-unrestricted CCSD calculations (UCCSD) based on restricted open-shell Hartree–Fock (ROHF) reference wave functions and semicanonical orbitals (ROHF-UCCSD level).
At the fc-CCSD/cc-pCVQZ level, not only the error in the electron-correlation contribution, but also the error in the Hartree-Fock contribution is relatively large (up to a few kJ/mol). In the previous study [108], these two errors were, to some extent, captured by the empirical factor. Within the present study, the unscaled CCSD(F12) energies are used as the correction to the basis set truncation error (*vide supra*), therefore the Hartree-Fock and (T) contributions were improved by increasing the basis set to the cc-pCV5Z [140]. The appropriate corrections are denoted with "δ_{HF}" and "(T)" in Table 7. The explicitly correlated corrections to the atomization energies (denoted with "F12" in Table 7) were obtained at the fc-CCSD(F12) level of theory with the def2-QZVPP orbital basis set [141]. For the density fitting approximation, the aug-cc-pwCV5Z MP2 fitting basis of Hättig was used (aug-cc-pV5Z for H) [93]. The incremental contribution denoted with "Other" is entirely taken from the previous study [108] and contains the following contributions

$$\Delta_{\mathrm{Other}} = \Delta_{\mathrm{CV}} + \Delta_{\mathrm{ZPVE}} + \Delta_{\mathrm{Anh.}} + \Delta_{\mathrm{MVD}} + \Delta_{\mathrm{SO}} + \Delta_{\mathrm{T}} + \Delta_{\mathrm{(Q)}}. \tag{271}$$

Δ_{CV} is a core-valence correction obtained as the difference between ae-CCSD(T)/cc-pCVQZ and fc-CCSD(T)/cc-pCVQZ energies. Δ_{ZPVE} is the harmonic zero-point vibrational correction obtained at the ae-CCSD(T)/cc-pCVTZ level, $\Delta_{\mathrm{Anh.}}$ is the correction due to anharmonic effects, calculated at the fc-MP2/cc-pVDZ level. Δ_{MVD} is the correction for scalar-relativistic effects (one electron Darwin and mass-velocity terms) obtained at the ae-CCSD(T)/cc-pCVTZ level [101, 102]. Δ_{SO} is a spin-orbit coupling correction, which may be non-zero only for open-shell species. For the C, O and F atoms, Δ_{SO} amounts to -0.35599, -0.93278 and -1.61153 kJ/mol, respectively [103]. The remaining contributions take care of the correction to the full triple excitations and perturbative treatment of quadruples $\Delta_{\mathrm{T}} = E_{\mathrm{CCSDT/cc\text{-}pVTZ}} - E_{\mathrm{CCSD(T)/cc\text{-}pVTZ}}$, $\Delta_{\mathrm{(Q)}} = E_{\mathrm{CCSDT(Q)/cc\text{-}pVDZ}} - E_{\mathrm{CCSDT/cc\text{-}pVDZ}}$. The final atomization energies are obtained by adding all the incremental contributions

$$\mathrm{AE}_{\mathrm{Total}} = \Delta_{\mathrm{CCSD}} + \delta_{\mathrm{HF}} + \Delta_{\mathrm{(T)}} + \Delta_{\mathrm{F12}} + \Delta_{\mathrm{Other}}, \tag{272}$$

and these values can be found in the column denoted by "Total". Such accurate treatment allows for a direct comparison with the experimental values. In the column "ATcT", the thermochemical data are presented. These values were taken from the Active Thermochemical Tables (ATcT) [109, 110, 111, 112] and they are considered as reliable and internally consistent values. As opposed to the traditional, sequential approach, the ATcT derive their results from a thermochemical network (TN) using all available knowledge. The thermochemical data used in this work were obtained from the latest version of the Core (Argonne) Thermochemical Network, C(A)TN, which is currently under development [142].

6.3 Computational Details

The conventional coupled-cluster and second-order Møller-Plesset calculations were performed with the Mainz-Austin-Budapest 2005 version of the ACES II program [83] and with the MRCC program [89, 90]. The former program was employed for the conventional CCSD approach with the perturbative correction for connected triple excitations (T). The MRCC code was used for the higher order coupled-cluster treatment, *i.e.* the full triples (CCSDT) and perturbative quadruples treatment (CCSDT(Q)). The CCSD(F12) method, as implemented in the TURBOMOLE program, was used for the purpose of the explicitly correlated calculations.
The equilibrium geometries of all considered molecules were obtained in the previous study [108] at the all-electron CCSD(T) level in the correlation consistent core-valence triple-ζ basis set (cc-pCVTZ) of Dunning [95] and Woon and Dunning [140]. For the hydrogens contained in the molecules, the cc-pVTZ basis set was used [95]. All the molecules are closed-shell species, therefore the RHF reference wave function was used as the reference for the geometry relaxation. The equilibrium structures are presented in Figures 8, 9, 10, 11 and 12.
The anharmonic correction, taken from Ref. [108], was obtained at the fc-MP2 level of theory with the cc-pVDZ basis set [143]. The cubic force field and those parts of the quartic force field that are required for the determination of the anharmonic effects were obtained by means of the numerical differentiation of the analytical Hessian around the equilibrium structure, as implemented in the ACES II program [83, 144]. The harmonic ZPVEs were obtained at the same level as the equilibrium geometries, *i.e.* ae-CCSD(T)/cc-pCVTZ, and the anharmonic corrections were computed at the fc-MP2/cc-pVDZ level. This is the only contribution that is obtained at a different equilibrium geometry than the other corrections.
In the column denoted by "CCSD" in Table 7, the ae-CCSD/cc-pCVQZ energies are presented. These values, taken from the study of Klopper *et al.* [108], are shown for the comparison with the previous results. Together with the column "Other" they can be considered as the "starting point" for the improvements carried out in the present study. The core-valence correction Δ_{CV}, contained in the formula Eq. (271), is computed as the difference between the ae-CCSD(T)/cc-pCVQZ and fc-CCSD(T)/cc-pCVQZ energies, where for the treatment of the hydrogens the cc-pVQZ basis set was used. This corrections is obtained using the ACES II program and, for the open-shell atoms, the ROHF reference employing the semi-canonical orbitals was used [145]. The correction Δ_T was obtained using the cc-pVTZ basis set [143] as the difference between the fc-CCSDT and fc-CCSD(T) energies. The perturbative quadruples correction $\Delta_{(Q)}$ was obtained within the cc-pVDZ basis set [143] and it should be considered as the incremental contribution

with respect to the CCSDT values. The higher order coupled-cluster calculations (CCSDT and CCSDT(Q)) were carried out with the MRCC program [89] and the open-shell atoms were treated with the UHF reference wave function.

The main purpose of the present study was to obtain the atomization energies of comparable quality to those calculated by Klopper *et al.* [108] without using the empirical "interference" factor. There are three major modifications of the current additivity scheme with respect to the one used in Ref. [108]. First, the Hartree-Fock component of the atomizations energies, previously computed with the cc-pCVQZ basis set [143], now is computed with the cc-pCV5Z basis set [140]. The appropriate incremental values, computed as the difference between the HF/cc-pCVQZ and HF/cc-pCV5Z energies, can by found in the column "δ_{HF}" of Table 7. Second, it has been discovered that the contribution from the perturbative triple excitation (T), obtained before with the cc-pCVQZ basis set, also introduces a non-negligible error to the final values. Therefore an additional correction, denoted with "(T)", has been included. This correction is the (T) contribution obtained with the cc-pCV5Z basis set [140]. The third, most important modification, changes the manner, in which the correction to the basis set truncation error is calculated. In the previous study, the explicitly correlated second order Møller-Plesset (MP2-F12) perturbation theory was applied. To get agreement with experimental values, the authors introduced an empirical scaling factor. This scaling factor was optimized against the mean deviation from the ATcT values. The F12 contribution to the atomization energies was weighted with the scaling factor $f_{\mathrm{int}} = 0.78$ and the final atomization energies were in very good agreement with the experimental values. Within the present work, instead of using the MP2-F12 values, the CCSD(F12) contributions are adopted as the corrections to the basis set truncation error. The CCSD(F12) calculations were carried out using the implementation in the TURBOMOLE program. The Slater-type correlation factor was used [Eq. (60)] with $\gamma = 1.4\ a_0^{-1}$, represented as a linear combination of six Gaussian functions with exponents and linear coefficients taken from Ref. [44]. The details associated with the theory and implementation of this method are presented in the thesis (see Sections 3 and 4). It is important to note that three variants of the explicitly correlated coupled-cluster method were used, the full CCSD(F12) model with optimized $c_{i,j}^{x,y}$ coefficients, and two approximated models, the fixed-amplitudes procedure proposed by Tew *et al.* [60] (called CCSD(F12/fixed)), and the model CCSD-F12b*, which refers to the work of Adler *et al.* [61, 62] (see Subsection 4.9 for the details about the implementation of these models).

6.4 Results and discussion

The components of the atomization energies are presented in Table 7. Where possible, they are compared to the ATcT data. In the last column, denoted with "Error", the deviations between the theoretical and experimental atomization energies are shown. The largest deviations, 7.1 and 6.7 kJ/mol, refer to the molecules 5 (tetrafluoromethane) and 83 (perfluoroperoxide), respectively. The reason for such significant discrepancy can be associated with the multireference character of these molecules. The D1 diagnostic of Jansen and Nielsen [146], obtained for these molecules, amounts to, respectively, 0.027 and 0.048. Both values exceed 0.02 which is the accepted threshold above which the electronic ground state is not well described by a single determinantal reference wave function. In Figure 7 the correlation between the F12 corrections obtained at the MP2-F12 and CCSD(F12) levels of theory is shown. The former quantities were weighted with the empirical interference factor $f_{\mathrm{int}} = 0.78$ [108]. The coupled-cluster values remain unscaled, but they contain the correction associated with the improvements made in

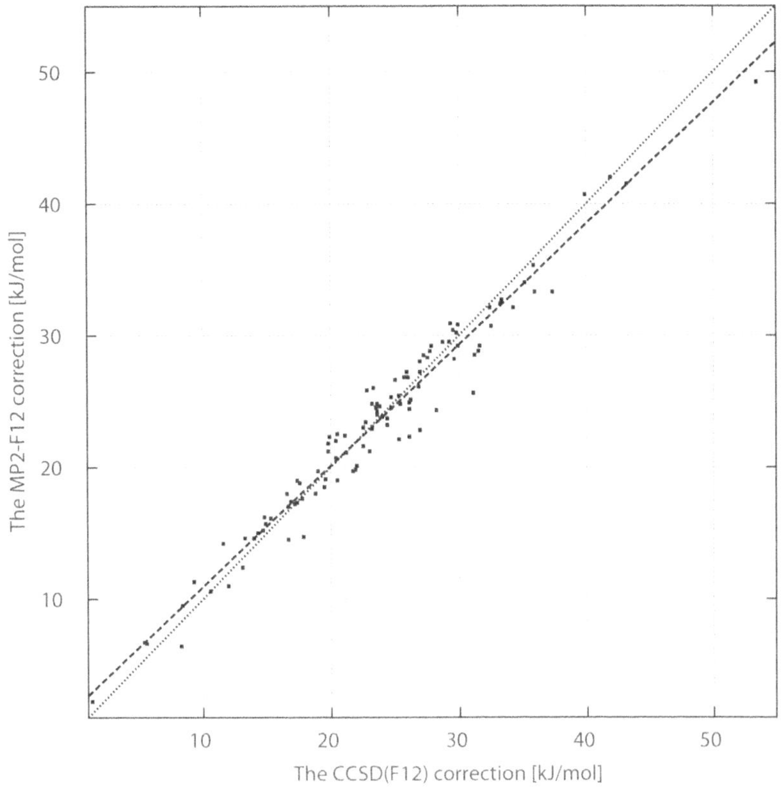

Figure 7: The correlation between the F12 contributions obtained at the CCSD(F12) and MP2-F12 levels of theory. The MP2-F12 values were scaled with the interference factor $f_{\text{int}} = 0.78$ determined by minimizing the deviations between the calculated and experimental atomization energies. The CCSD(F12) values contain the correction associated with the improvements made in the (T) contribution. The dashed line refers to the linear function fitted to the presented data and the dotted line reflects the hypothetical perfect correlation.

the (T) contribution. The dashed line refers to the linear function fitted to the data points on the plot, the dotted line reflects the hypothetical perfect correlation between the CCSD(F12) and MP2-F12 quantities. These contributions correlate well, the deviation between the dashed and dotted lines is rather small and amounts to 1.10 kJ/mol, in terms of the standard deviation of the data points from the dotted line. This means that the choice of the empirical scaling factor, in the case of the MP2-F12 results, was correct and efficient. It was possible to enhance the MP2-F12 contributions to the values that are close to the CCSD(F12) results with just one factor. The remaining differences, *i.e.* the deviations between the dashed and dotted line on Figure 7, refer to the fact that the factor was chosen to minimize the deviation from the experimental rather than from the CCSD(F12) values. It is likely that it also corrects the contributions that go beyond the CCSD(T) level. The removal of the scaling factor, in the case of the F12 corrections based on CCSD(F12) results, required the improvements of the Hartree-Fock and perturbative triples (T) components of the atomization energies. The results collected in Table 7 are statistically analyzed in Table 6.

Table 6: Statistics of the deviations of the computed values from the ATcT reference data (all deviations in kJ/mol).

F12 contribution	δ_{ave} [a]	δ_{mad} [b]	δ_{rms} [c]	95% [d]	δ_{max} [e]	Molecule[f]
			Errors per molecule			
CCSD(F12)	0.03	1.69	2.19	4.39	7.1	CF_4
CCSD(F12/fixed)	1.99	2.56	3.38	6.76	10.3	CF_4
CCSD-F12b*	−2.47	2.85	3.31	6.62	−7.9	C_3H_6 (cyclopropane)
MP2-F12(f_{int} = 0.0)[g]	−21.7	21.7	23.1	46.3	−52.4	N_2O_4
MP2-F12(f_{int} = 0.78)[g]	−0.12	0.90	1.22	2.44	4.1	C_2H_3F
MP2-F12(f_{int} = 1.0)[g]	5.98	5.98	6.52	13.0	14.1	N_2O_4
			Without improving Hartree-Fock and (T)			
CCSD(F12)	−2.10	2.45	2.79	5.58	−6.6	N_2H_4
CCSD(F12/fixed)	−0.14	1.73	2.21	4.43	7.5	CF_4
CCSD-F12b*	−4.60	4.60	5.14	10.27	−0.4	C_3H_6
			Errors per valence electron			
CCSD(F12)	−0.03	0.11	0.13	0.27	−0.4	NH_3
CCSD(F12/fixed)	0.09	0.14	0.17	0.34	0.4	CO_2
CCSD-F12b*	−0.17	0.19	0.23	0.46	−0.6	NH_3
MP2-F12(f_{int} = 0.0)[g]	−1.33	1.33	1.36	2.72	−1.9	H_4N_2
MP2-F12(f_{int} = 0.78)[g]	−0.01	0.06	0.08	0.16	0.2	C_2H_3F
MP2-F12(f_{int} = 1.0)[g]	0.37	0.37	0.38	0.77	0.7	H_2N_2 (cis-diazene)
			Without improving Hartree-Fock and (T)			
CCSD(F12)	−0.15	0.17	0.20	0.40	−0.5	NH_3
CCSD(F12/fixed)	−0.04	0.11	0.15	0.29	−0.4	NH_3
CCSD-F12b*	−0.30	0.30	0.34	0.67	−0.7	NH_3

[a] Mean error.
[b] Mean absolute error.
[c] Root-mean-square error.
[d] 95% confidence limit.
[e] Maximum deviation.
[f] Molecule with largest error.
[g] Data taken from Ref. [108].

The standard mean error amounts to ±0.13 kJ/mol per valence electron when the F12 correction based on the CCSD(F12) calculations is included. The F12 corrections based on the alternative approximations perform slightly worse, the standard mean errors are ±0.17 and ±0.23 kJ/mol per valence electron, respectively, for the CCSD(F12/fixed) and CCSD-F12b* models. In terms of the errors per molecule, the difference between the RMS errors of full CCSD(F12) and the other models is ca. 1 kJ/mol. This suggests that it is preferable to use the F12 contributions based on the full CCSD(F12) calculations. The performance of the additivity scheme based on the scaled MP2-F12 contributions is still slightly better. The RMS deviation per valence electron is lower by ca. 0.05 kJ/mol and amounts to 0.08 kJ/mol per valence electron. The reason for this is probably associated with the fact that the interference scaling factor accounts not only for the effects at the CCSD level. If that were the case, then the scaled MP2-F12 and CCSD(F12) results should be the same. The usage of the unscaled CCSD(F12) rather than scaled MP2-F12 contributions without modifying the rest of the procedure led to a worsening of the statistics (see Table 6 under the label "Without improving Hartree-Fock and (T)").

The root mean square error per molecule is ca. 0.6 kJ/mol larger when the Hartree-Fock and (T) components were not improved. This effect is even more pronounced for the CCSD-F12b* approximation, where this difference is ca. 2.8 kJ/mol. In the case of the CCSD(F12/fixed) approximation, one observes significant improvement of the statistics, which is likely due to a fortunate cancellation of the errors.
The inclusion of the unscaled CCSD(F12) contribution as the correction to the basis set truncation error led to the conclusion that the presence of the scaling factor corrects also the errors coming from other sources. Therefore it was a natural idea to improve the (T) and Hartree-Fock components of the atomization energies, and new terms were taken into account (fifth and sixth column in Table 7). The authors in Ref. [108] suggest that the remaining error, appearing in their additivity scheme, amounts to $\pm$0.13 kJ/mol per valence electron. This is due to the propagation of the errors associated with the following contributions (per valence electron): Hartree-Fock ($\pm$0.08 kJ/mol), core-valence ($\pm$0.04 kJ/mol), ZPVE ($\pm$0.06 kJ/mol), relativistic effects ($\pm$0.003 kJ/mol), full triples and (Q) ($\pm$0.05 kJ/mol), valence-shell post-CCSDT(Q) ($\pm$0.02 kJ/mol), core-valence post-CCSD(T) ($\pm$0.03) and DBOC ($\pm$0.02 kJ/mol). The improvements done within the present study, *i.e.* the extension of the basis set at the Hartree-Fock and the (T) levels, reduce this value to ca. 0.10 kJ/mol. The final atomization energies, plagued with such small error, were obtained without using the interference scaling factor.

6.5 Conclusions

The atomization energies of 105 molecules from the test set of Bakowies [139] have been computed with an estimated standard deviations from the ATcT values of $\pm$0.13 kJ/mol per electron in the valence shell. This agreement was achieved by including the CCSD(F12) contribution that corrects for the basis set truncation error. It was possible to achieve good agreement with the experimental values without falling back on empirical corrections or extrapolations. In comparison to the previous study of Klopper *et al.* [108], the scaled MP2-F12 contributions were replaced with unscaled CCSD(F12) values. Together with improving the picture of the Hartree-Fock and (T) components, the final atomization energies, obtained without the empirical scaling, are of similar accuracy to those from the study of Klopper *et al.*. The estimated error of the presented additivity scheme amounts to 0.10 kJ/mol per valence electron.

Table 7: Atomization energies in kJ/mol.[a]

Nr.[b]	Molecule[c]		CCSD	δ_{HF}	(T)	F12	Other[d]	Total	ATcT	Error
1	CFN	cyanogen fluoride	1202.7	1.1	50.8	17.9	−21.0	1251.6	1250.3 ± 1.7	1.3
2	CFN	isocyanogen fluoride	902.4	1.2	54.5	16.8	−17.2	957.8	959.2 ± 2.7	−1.4
3	CF_2	singlet difluoromethylene	1028.4	1.4	37.8	14.7	−20.0	1062.2	1059.1 ± 0.8	3.1
4	CF_2O	carbonyl fluoride	1677.5	1.6	56.0	23.6	−37.7	1721.0	1718.4 ± 0.9	2.6
5	CF_4	tetrafluoromethane	1922.1	1.9	54.4	28.3	−51.7	1955.0	1947.9 ± 0.6	7.1
6	CHF	singlet fluoromethylene	855.1	0.9	23.3	9.3	−31.7	857.0		
7	$CHFO$	formyl fluoride	1620.0	1.2	45.6	18.6	−52.9	1632.4	1631.4 ± 0.9	1.0
8	CHF_3	trifluoromethane	1855.1	1.5	43.9	23.0	−71.1	1852.4	1848.7 ± 0.9	3.7
9	CHN	hydrogen cyanide	1250.1	0.7	37.9	14.1	−35.1	1267.7	1268.3 ± 0.2	−0.6
10	CHN	hydrogen isocyanide	1189.6	0.8	36.2	13.6	−35.4	1204.9	1207.0 ± 0.6	−2.1
11	$CHNO$	cyanic acid	1627.8	1.3	56.1	21.8	−49.7	1657.2	1657.2 ± 1.0	0.0
12	$CHNO$	isocyanic acid	1725.3	1.3	60.3	22.5	−48.7	1760.7	1761.0 ± 0.4	−0.3
13	$CHNO$	formonitrile oxide	1418.5	1.2	73.2	21.6	−41.4	1473.1	1474.1 ± 1.2	−1.0
14	$CHNO$	isofulminic acid	1377.6	1.4	58.0	20.9	−48.1	1409.8	1410.2 ± 1.0	−0.4
15	CH_2	singlet methylene	742.1	0.5	7.8	4.7	−41.0	714.0	714.9 ± 0.2	−0.9
16	CH_2F_2	difluoromethane	1778.3	1.1	32.8	17.7	−87.3	1742.6	1741.7 ± 0.8	0.9
17	CH_2N_2	cyanamide	1937.9	1.4	56.7	24.5	−80.1	1940.3		
18	CH_2N_2	3H-diazirine	1754.3	1.7	63.7	24.2	−79.9	1763.9		
19	CH_2N_2	diazomethane	1788.7	1.4	66.9	23.3	−71.5	1808.9		
20	CH_2O	formaldehyde	1515.0	0.8	32.6	13.4	−65.7	1496.1	1495.8 ± 0.2	0.3
21	CH_2O	hydroxymethylene	1300.8	1.0	30.0	12.8	−67.3	1277.3	1277.8 ± 1.1	−0.5
22	CH_2O_2	dioxirane	1627.0	1.6	60.5	21.6	−81.6	1629.0	1629.6 ± 1.7	−0.6
23	CH_2O_2	formic acid	2019.2	1.3	51.8	22.1	−85.0	2009.3	2008.4 ± 0.3	0.9
24	CH_2O_3	performic acid	2133.8	1.9	77.3	29.2	−91.4	2150.7		
25	CH_3F	fluoromethane	1729.6	0.8	21.8	12.7	−102.0	1662.8	1665.1 ± 0.6	−2.3
26	CH_3N	methanimine	1778.4	1.0	34.9	15.8	−98.2	1731.9	1733.5 ± 1.0	−1.6
27	CH_3NO	formamide	2288.1	1.5	52.9	24.9	−111.6	2255.7		
28	CH_3NO_2	methyl nitrite	2379.7	2.1	86.7	31.5	−119.9	2380.1		
29	CH_3NO_2	nitromethane	2385.0	2.1	89.3	32.1	−122.8	2385.8		
30	CH_4	methane	1733.2	0.5	11.9	7.7	−113.6	1639.7	1642.2 ± 0.1	−2.5
31	CH_4N_2O	urea	3029.4	2.1	71.0	36.3	−158.5	2980.2		
32	CH_4O	methanol	2096.9	1.0	28.2	15.9	−130.9	2011.1	2012.7 ± 0.2	−1.6
33	CH_5N	methylamine	2378.0	1.1	30.0	18.2	−161.2	2266.0	2269.0 ± 0.5	−3.0
34	CO	carbon monoxide	1038.8	0.7	33.2	10.9	−10.1	1073.5	1072.1 ± 0.1	1.4
35	CO_2	carbon dioxide	1548.6	1.2	57.3	19.8	−25.8	1601.1	1598.2 ± 0.1	2.9
36	C_2F_2	difluoroacetylene	1526.3	1.2	56.5	20.6	−29.0	1575.7	1577.0 ± 1.7	−1.3
37	C_2F_4	tetrafluoroethylene	2350.1	2.2	76.6	33.9	−56.7	2406.2	2405.2 ± 1.0	1.0
38	C_2HF	fluoroacetylene	1593.4	0.9	45.7	16.3	−45.5	1610.8	1612.3 ± 1.0	−1.5
39	C_2HF_3	trifluoroethylene	2345.6	1.8	64.9	28.7	−75.0	2365.9		
40	C_2H_2	acetylene	1637.4	0.5	34.4	12.3	−60.4	1624.3	1626.2 ± 0.2	−1.9
41	$C_2H_2F_2$	1,1-difluoroethylene	2367.7	1.5	53.4	23.7	−92.0	2354.3		
42	C_2H_2O	ketene	2145.9	1.1	53.8	19.4	−73.4	2146.9	2147.3 ± 0.2	−0.4
43	C_2H_2O	oxirene	1813.6	1.4	59.2	21.1	−67.7	1827.6		
44	$C_2H_2O_2$	glyoxal	2547.3	1.6	70.7	27.1	−89.0	2557.8	2555.3 ± 0.6	2.5
45	C_2H_3F	fluoroethylene	2329.1	1.1	42.0	18.6	−108.6	2282.2	2278.4 ± 1.7	3.8
46	C_2H_3FO	acetyl fluoride	2860.0	1.6	60.4	27.0	−122.5	2826.5		
47	C_2H_3N	acetonitrile	2491.3	1.1	52.3	21.6	−108.2	2458.2		
48	C_2H_3N	methyl isocyanide	2391.4	1.4	50.7	21.4	−109.9	2355.0		
49	C_2H_4	ethylene	2304.3	0.8	30.5	13.5	−124.2	2224.8	2225.9 ± 0.2	−1.1
50	C_2H_4O	acetaldehyde	2755.0	1.3	47.9	21.4	−137.2	2688.3	2688.9 ± 0.4	−0.6
51	C_2H_4O	oxirane	2643.5	1.5	48.5	22.8	−143.4	2572.9	2573.9 ± 0.5	−1.0
52	$C_2H_4O_2$	acetic acid	3254.9	1.7	66.5	30.4	−154.3	3199.1	3199.3 ± 1.5	−0.2
53	$C_2H_4O_2$	methyl formate	3183.6	1.8	67.5	30.2	−155.8	3127.3	3125.2 ± 0.6	2.1

Nr.[b]	Molecule[c]		CCSD	δ_{HF}	(T)	F12	Other[d]	Total	ATcT	Error
54	C_2H_5F	fluoroethane	2951.0	1.2	36.8	20.5	−172.1	2837.4	2838.5 ± 1.9	−1.1
55	C_2H_5N	aziridine	2925.7	1.6	49.8	25.1	−174.6	2827.6		
56	C_2H_6	ethane	2930.7	0.9	26.2	15.3	−188.6	2784.6	2787.2 ± 0.2	−2.6
57	C_2H_6O	dimethyl ether	3264.5	1.5	43.5	23.7	−201.4	3131.7	3132.4 ± 0.5	−0.7
58	C_2H_6O	ethanol	3315.0	1.4	43.4	23.6	−201.7	3181.7	3182.8 ± 0.3	−1.1
59	C_2N_2	cyanogen	1970.9	1.4	84.8	26.7	−27.5	2056.1	2055.8 ± 0.5	0.3
60	C_3H_3N	acrylonitrile	3071.4	1.5	73.7	27.3	−117.1	3056.8		
61	C_3H_4	allene	2858.6	1.0	50.5	19.2	−131.7	2797.6	2800.9 ± 0.5	−3.3
62	C_3H_4	cyclopropene	2765.9	1.2	50.5	20.7	−134.6	2703.7	2705.1 ± 1.0	−1.4
63	C_3H_4	propyne	2868.0	1.1	49.2	17.3	−132.1	2803.4	2805.6 ± 0.5	−2.2
64	C_3H_6	cyclopropane	3490.0	1.4	44.5	22.5	−203.0	3355.4	3359.7 ± 0.6	−4.3
65	C_3H_6	propene	3520.5	1.2	45.9	21.2	−195.3	3393.6	3395.0 ± 0.4	−1.4
66	C_3H_8	propane	4136.2	1.3	41.5	23.0	−257.7	3944.4	3944.6 ± 0.4	−0.2
67	C_3O_2	carbon suboxide	2620.5	1.7	110.9	31.3	−41.0	2723.5		
68	C_4H_4	butatriene	3415.2	1.3	75.2	24.8	−137.8	3378.7		
69	C_4H_4	cyclobutadiene	3308.7	1.6	76.1	26.6	−142.3	3270.7		
70	C_4H_4	tetrahedran	3200.0	1.7	69.8	29.4	−140.3	3160.6		
71	C_4H_4	vinylacetylene	3454.5	1.3	70.6	25.5	−141.7	3410.3		
72	C_4N_2	dicyanoacetylene	3132.7	1.9	127.9	38.3	−44.8	3255.9		
73	FH	hydrogen fluoride	577.0	0.3	8.7	4.9	−26.0	564.9	566.0 ± 0.0	−1.1
74	FHO	hypofluorous acid	616.2	0.9	33.4	11.4	−36.8	625.1	624.0 ± 0.1	1.1
75	FHO_2	fluoroperoxide	793.6	1.6	61.6	19.0	−46.5	829.4		
76	FH_2N	monofluoroamine	1023.9	1.1	29.9	14.9	−71.4	998.3		
77	FH_3N_2	fluorohydrazine	1628.0	1.8	52.1	26.3	−115.0	1593.2		
78	FNO	nitrosyl fluoride	806.7	1.8	68.2	19.9	−16.0	880.5		
79	F_2	difluorine	123.1	0.9	31.2	6.8	−6.3	155.8	154.6 ± 0.2	1.2
80	F_2N_2	difluorodiazene (cis)	941.9	2.2	79.6	27.7	−29.0	1022.4		
81	F_2N_2	difluorodiazene (trans)	938.7	2.4	76.7	27.2	−28.8	1016.2		
82	F_2O	difluorine monoxide	317.8	1.8	56.3	14.9	−14.1	376.7	373.3 ± 0.7	3.4
83	F_2O_2	perfluoroperoxide	505.4	2.3	102.4	22.4	−16.1	616.4	609.7 ± 0.8	6.7
84	F_3N	trifluoroamine	771.3	2.2	65.2	24.5	−30.4	832.8		
85	HNO	nitrosylhydride	800.0	1.2	41.5	14.7	−33.4	824.0	823.6 ± 0.1	0.4
86	HNO_2	nitrous acid (cis)	1208.8	1.7	68.7	23.2	−49.5	1252.9	1251.5 ± 0.4	1.4
87	HNO_2	nitrous acid (trans)	1209.1	1.9	69.0	23.1	−49.0	1254.1	1253.3 ± 0.1	0.8
88	HNO_2	nitrous acid, H–NO_2	1170.4	1.8	74.6	23.7	−53.4	1217.0		
89	HNO_3	nitric acid	1491.0	2.2	95.9	32.0	−66.5	1554.6	1551.6 ± 0.2	3.0
90	HN_3	hydrogen azide	1272.7	1.8	76.9	25.8	−47.2	1330.0	1329.7 ± 0.6	0.3
91	H_2N_2	diazene (cis)	1151.4	1.4	41.8	17.6	−68.3	1143.9	1143.5 ± 0.9	0.4
92	H_2N_2	diazene (trans)	1174.0	1.3	41.0	17.7	−69.8	1164.2	1165.8 ± 0.7	−1.6
93	H_2N_2	diazene (iso)	1072.1	1.2	37.9	18.4	−65.5	1064.1	1065.1 ± 0.9	−1.0
94	H_2N_2O	nitrosamide	1515.2	1.8	66.6	26.6	−79.2	1530.9		
95	H_2O	water	948.0	0.5	14.4	8.2	−55.5	915.7	917.8 ± 0.1	−2.1
96	H_2O_2	hydrogen peroxide	1070.3	1.1	36.9	15.3	−68.5	1055.0	1055.2 ± 0.1	−0.2
97	H_3N	ammonia	1214.6	0.7	15.9	10.5	−87.3	1154.3	1157.3 ± 0.1	−3.0
98	H_3NO	ammonia oxide	1336.5	1.5	35.2	19.9	−105.7	1287.5		
99	H_3NO	hydroxylamine	1446.1	1.2	34.8	18.3	−103.3	1397.1	1398.7 ± 0.5	−1.6
100	H_4N_2	hydrazine	1769.1	1.3	34.7	21.0	−135.1	1691.2	1695.6 ± 0.2	−4.4
101	N_2	dinitrogen	894.5	1.1	39.1	15.5	−9.3	940.8	941.1 ± 0.1	−0.3
102	N_2O	nitrous oxide	1023.6	1.6	77.7	22.8	−22.3	1103.4	1102.0 ± 0.1	1.4
103	N_2O_3	dinitrogen trioxide	1445.7	2.8	144.1	38.3	−36.8	1594.2	1591.1 ± 0.2	3.1
104	N_2O_4	dinitrogen tetraoxide	1737.9	3.4	177.4	47.4	−53.9	1912.3	1908.5 ± 0.2	3.8
105	O_3	ozone	477.9	2.1	106.7	18.3	−8.0	597.0	596.1 ± 0.1	0.9
106	H_2	dihydrogen	456.6	0.1	0.0	1.2	−26.1	431.8	432.1 ± 0.0	−0.3

[a] The individual contributions are explained in the text.
[b] Same number and same molecule as in Ref. [139] except for dihydrogen.
[c] The ae-CCSD(T)/cc-pCVTZ equilibrium geometry was used for each molecule.
[d] Further corrections taken from Ref. [108]. See the text.

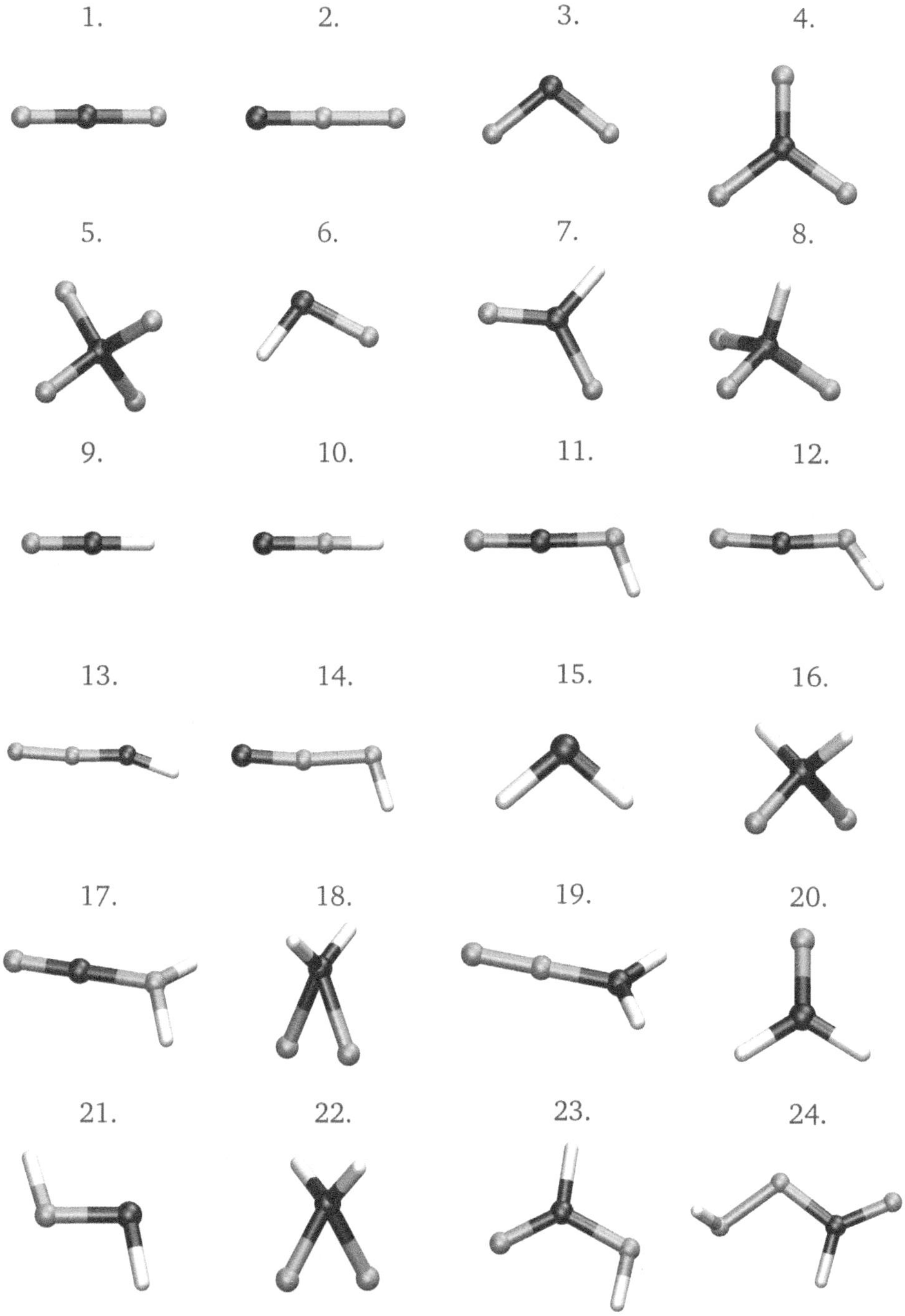

Figure 8: The structures of molecules 1 – 24 considered in Section 6. The geometries were obtained at the ae-CCSD(T) level of theory with the cc-pCVTZ basis set.

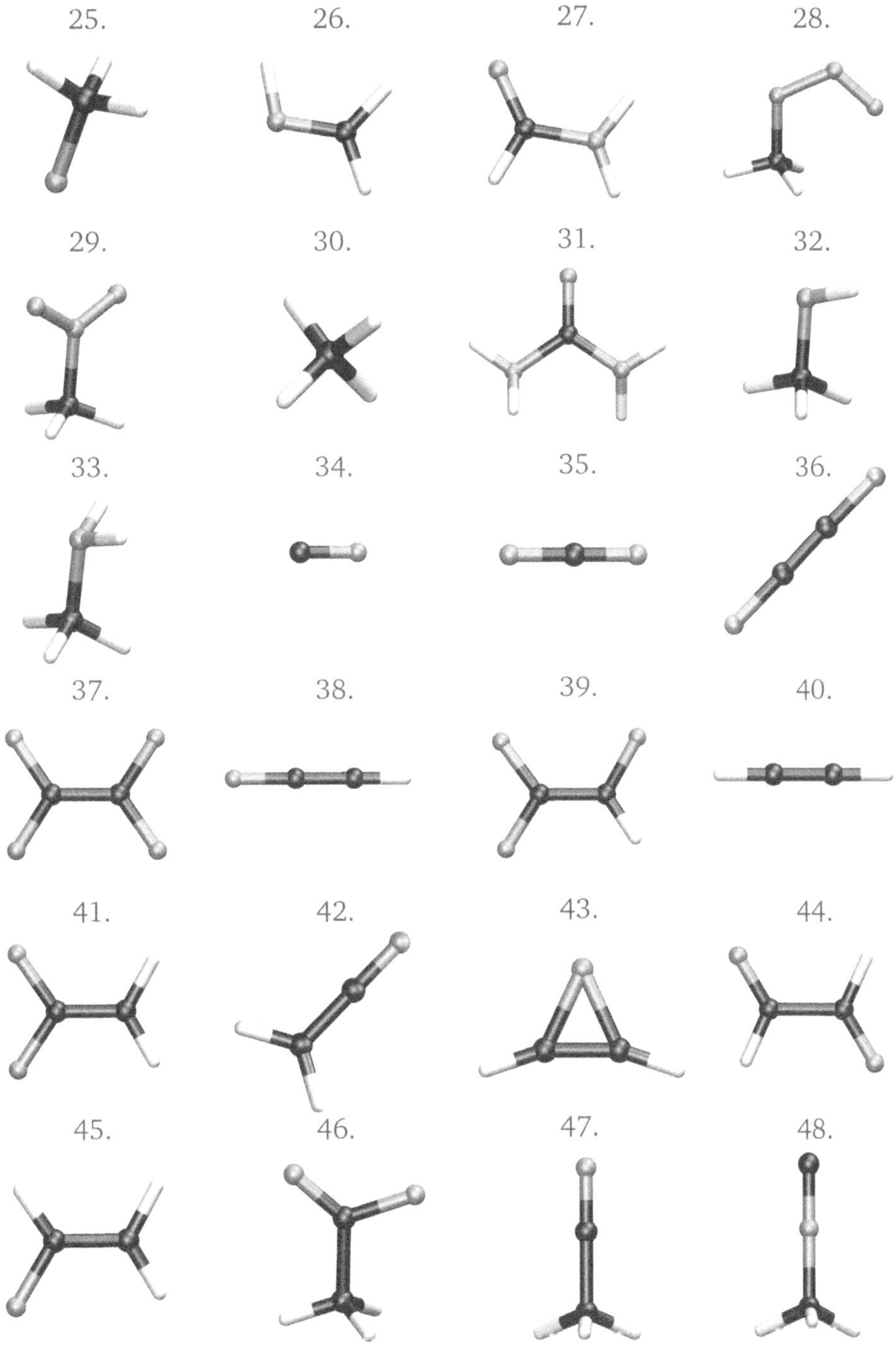

Figure 9: The structures of molecules 25 – 48 considered in Section 6. The geometries were obtained at the ae-CCSD(T) level of theory with the cc-pCVTZ basis set.

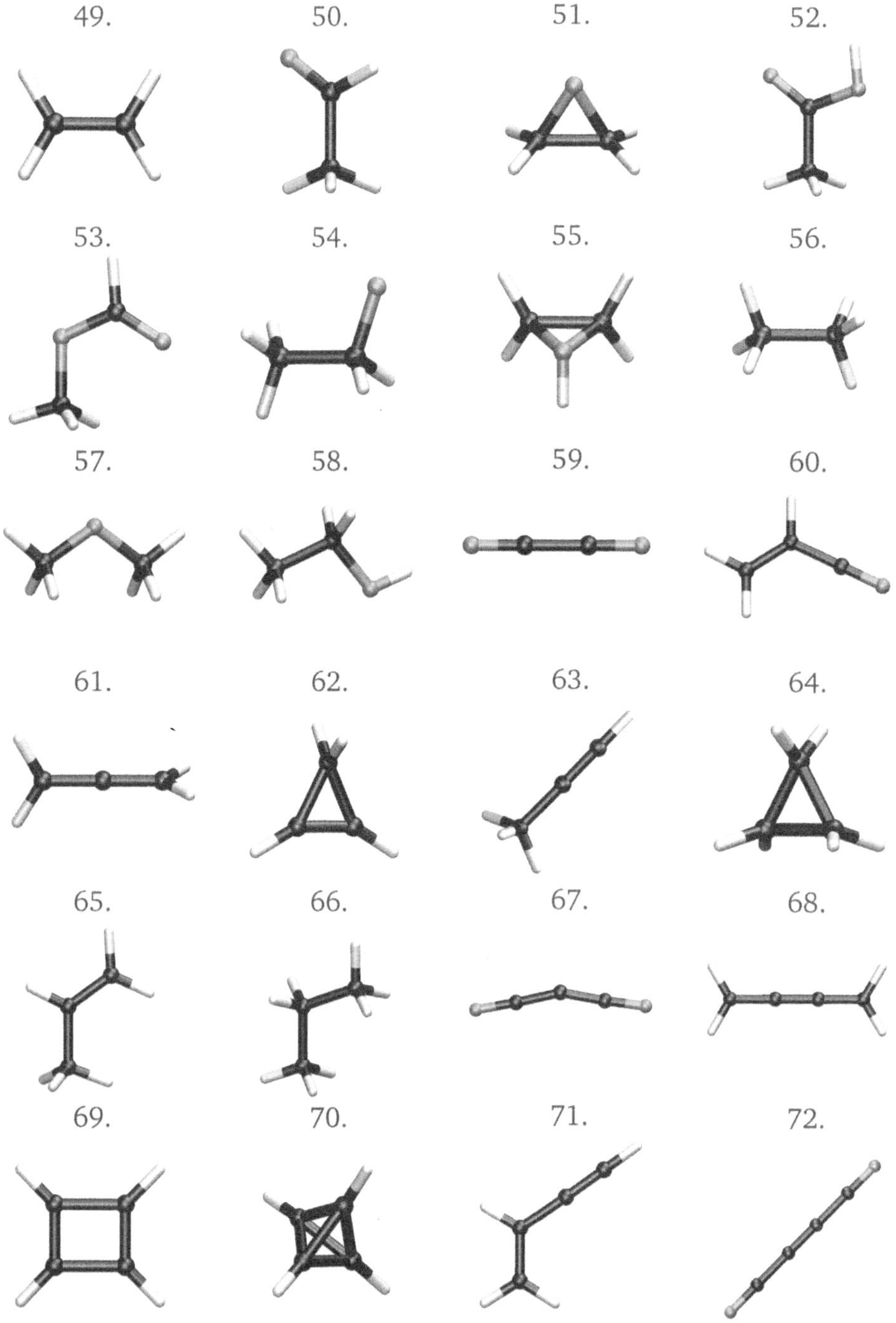

Figure 10: The structures of molecules 49 – 72 considered in Section 6. The geometries were obtained at the ae-CCSD(T) level of theory with the cc-pCVTZ basis set.

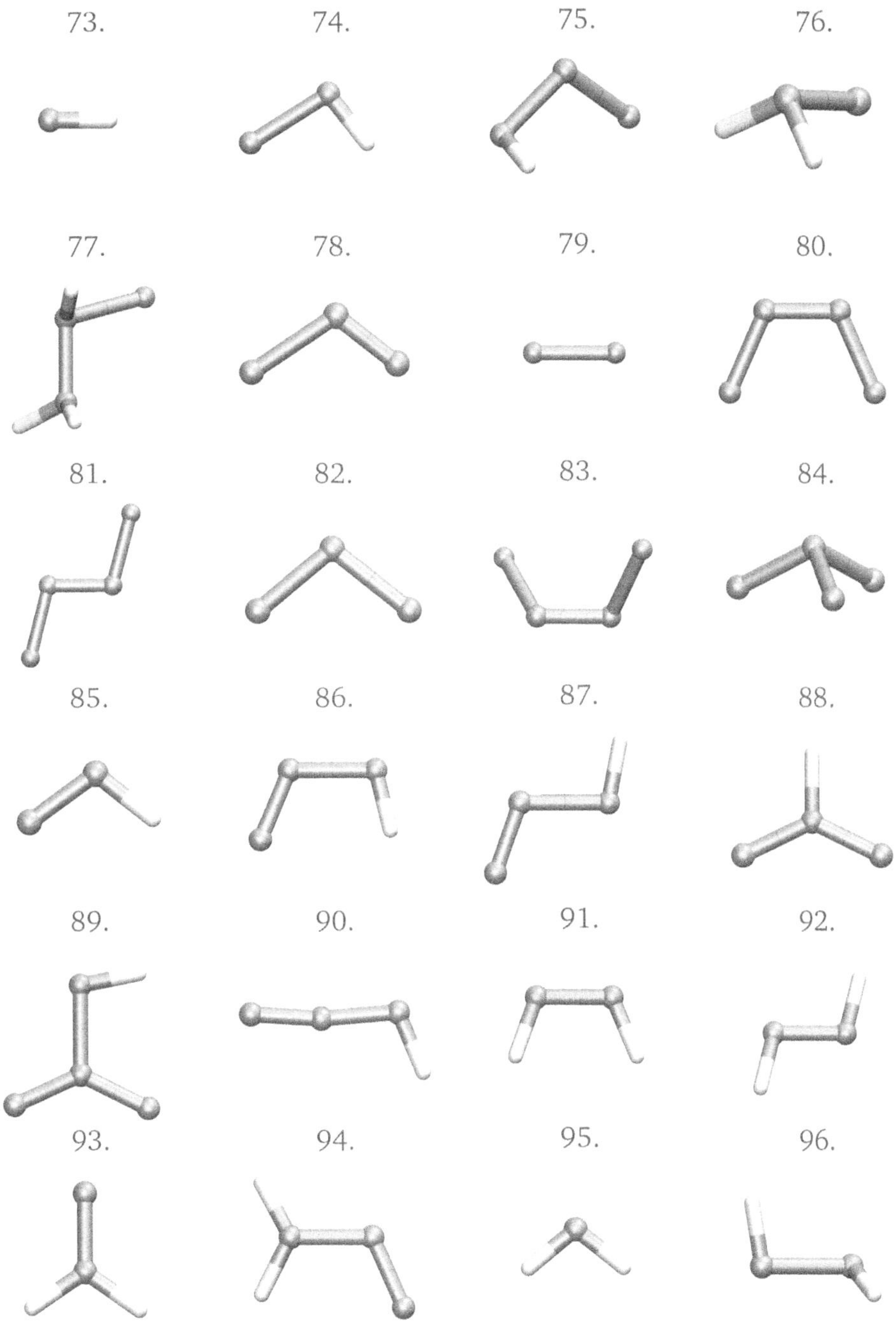

Figure 11: The structures of molecules 73 – 96 considered in Section 6. The geometries were obtained at the ae-CCSD(T) level of theory with the cc-pCVTZ basis set.

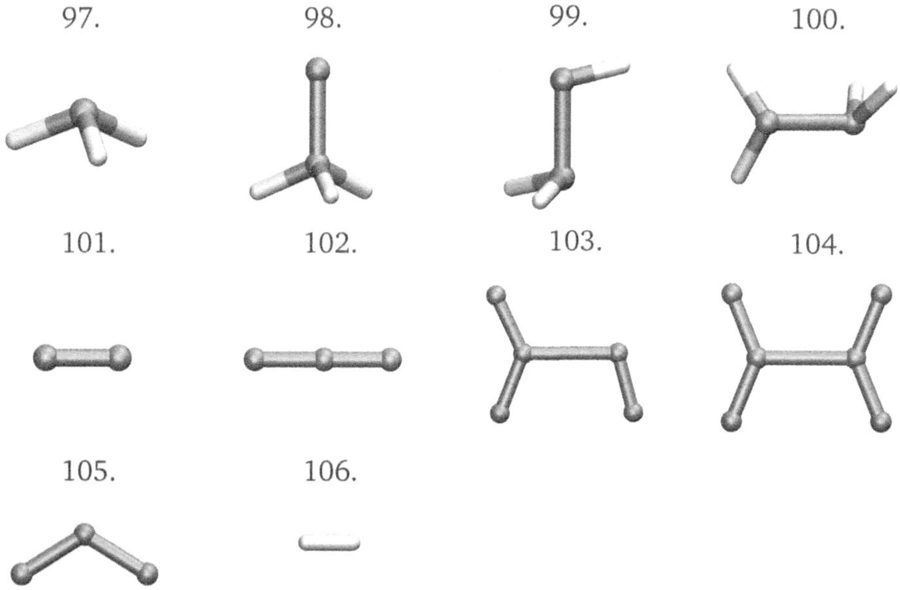

Figure 12: The structures of molecules 97 – 106 considered in Section 6. The geometries were obtained at the ae-CCSD(T) level of theory with the cc-pCVTZ basis set.

7 Coupled-cluster calculations of the ionization potentials and electron affinities of the atoms H, C, N, O and F

The ionization potentials and electron affinities of the atoms H, C, N, O and F have been computed by means of coupled-cluster methods using doubly augmented correlation-consistent one-electron basis sets in conjunction with explicitly correlated Slater-type geminals. Excitations up to the level of connected quintuples have been accounted for, and all orbitals in the core and valence shells have been correlated. Relativistic effects (spin–orbit as well as scalar) and diagonal Born–Oppenheimer corrections have been included.

7.1 Introduction

The last five years have seen a rapid development of explicitly correlated methods that use Slater-type geminals (STGs) as two-particle basis functions. In 2004, these STGs were introduced by Ten-no [45] into the explicitly correlated R12 methods of Kutzelnigg and co-workers [26, 147, 148, 149, 150, 151, 152, 153] thereby significantly advancing the accuracy and applicability of these methods. Also the introduction of auxiliary basis sets [47] to improve the resolution-of-the-identity (RI) approximations [147, 148, 149], which are intrinsic to the R12 methods, has initiated a renewed interest in the R12 methods and has led to the development of more efficient and more accurate approaches. When using R12 methods today, it is common practice to use STGs in combination with (complementary) auxiliary basis sets (CABS)[20].
In the following and throughout this Section, we shall refer to the R12 methods that use STGs as "F12 methods". These were first developed, investigated, and applied in the framework of second-order Møller–Plesset perturbation theory, that is, at the MP2-F12 level, evaluating the necessary integrals either by numerical quadrature or by representing the exponential with a linear combination of Gaussians [27, 44, 45, 154]. Numerical quadrature [24], robust density-fitting [21, 155] and/or local-correlation approaches have led to very powerful MP2-F12 implementations [23, 49, 50, 108, 156, 157, 158].
The main purpose of this study is the determination of the ionization potentials (IP) and electron affinities (AE) of the hydrogen, carbon, nitrogen, oxygen and fluorine atoms. In order to obtain accurate results, coupled-cluster techniques up to connected quintuple excitations have been combined with the CCSD(F12) method. The contributions obtained at the latter level of theory are considered as corrections to the basis set truncation error. For the purpose of the explicitly correlated treatment we have employed the integral-direct density-fitting implementation of the CCSD(F12) model, that recently became available as a part of the TURBOMOLE program [69]. Relativistic and diagonal Born-Oppenheimer corrections have been also included.

7.2 Computational details

All conventional coupled-cluster calculations were performed with the Mainz-Austin-Budapest version of the ACES II program [83] and with the MRCC program [89, 90]. Within all post Hartree-Fock calculations, both core and valence electrons were correlated (all-electron calculations). For the open-shell cases, the UHF reference wave function was used. The Hartree-Fock and CCSD(T) contributions to the IPs and EAs were computed within the correlation-consistent doubly augmented quintuple-ζ, d-aug-cc-pwCV5Z, basis set [97]. Incremental contributions referring to these levels of theory can be found under the label "CCSD(T)" in Tables 10 and 11, respectively for the IPs and EAs (relative quantities are presented in eV). The correction that accounts for the perturbative treatment of quadruple excitations with respect to the (T) values was computed with the d-aug-cc-pwCVQZ basis set [95]. This values are labeled with "(T)→(Q)" in Tables 10 and 11. Contributions associated with the incremental corrections from full quadruple excitations, denoted with "(Q)→Q", were computed with the d-aug-cc-pwCVTZ basis set [95]. They should be considered as the incremental corrections with respect to the (Q) values. The connected pentuple excitations were taken into account in a perturbative manner in the d-aug-cc-pwCVDZ basis set [95].

All explicitly correlated calculations were performed at the CCSD(F12) level of theory, as implemented in the TURBOMOLE program [58, 69]. The Slater-type correlation factor was used with the exponent $\gamma = 1.0\ a_0^{-1}$. It was approximated by a linear combination of six Gaussian functions with linear and nonlinear coefficients taken from Ref. [44]. The CCSD(F12) electronic energies were computed in an all-electron calculation with the d-aug-cc-pwCV5Z basis set [97]. For all cases we used full CCSD(F12) model (see Subsection 4.9 for the discussion about models implemented in TURBOMOLE), the open-shell species were computed with a UHF reference wave function. The explicitly correlated contributions to the relative quantities are collected in Tables 10 and 11 under the label "F12".

The mass-velocity and Darwin terms (denoted as "MVD") were computed at the ae-CCSD(T) level of theory [145] with the d-aug-cc-pwCV5Z basis set. The spin-orbit corrections ("SO") were obtained from the experimentally observed spin-orbit splitting [159]. The diagonal Born-Oppenheimer correction was was crudely estimated from the total electronic energy as

$$\Delta E_{\mathrm{DBOC}} = -\sum_I \frac{p_I}{M_I} E_{\mathrm{Total}}, \tag{273}$$

where the sum runs over the relevant isotopes with mass M_I and natural abundancy p_I. For the calculation of this correction we used the isotope relative atomic masses and abundances reported in Ref. [159]. The total electronic energies in Eq. (273) were obtained at the ae-CCSD(T)/d-aug-cc-pwCV5Z level.

For energies and relative atomic masses we used the following conversion factors 1 E_h = 27.21138386(68) eV and 1 u = 1.660538782(83) $\times 10^{-27}$ kg = 1822.88848 m_e, respectively, where m_e is the electron mass [160].

The aug-cc-pwCV5Z density-fitting [93] and aug-cc-pV5Z exchange-fitting [94] basis sets [the latter one used for the representation of the MP2-F12 $B_{x,y}^{m,n(ij)}$ intermediate, see Eq. (204)] have been supplemented with additional diffused Gaussian functions (Tables 8, 9). To test these basis sets, we have performed ae-CCSD(F12)/d-aug-cc-pwCV5Z calculations on the H^- and F^- anions with the DALTON program [161], using the same Ansatz (2B), same Slater-type geminal ($\gamma = 1.0\ a_0^{-1}$), and the same CABS (*vide supra*) as used with the TURBOMOLE program [69]. Within the implementation of the CCSD(F12) model in DALTON the DF techniques are not employed and all integrals are computed in an exact manner [31, 32]. DALTON yielded −0.5277231 E_h and −99.8469203 E_h for H^- and F^-, respectively, while TURBOMOLE yielded −0.5277227 E_h and −99.8469450 E_h. The density fitting error is of the order of magnitude of 2.5 μE_h in the case of fluorine anion, and it is expected to be much smaller in terms of relative quantities.

7.3 Results

The atomic ionization potentials and electron affinities are presented in Tables 10 and 11, respectively. In the first eight columns the incremental contributions to the final quantities are given, in the column labeled with "Calc." the sum of them is shown. The last column, denoted with "Exptl.", contains the experimental results. All values are provided in eV units. In the case of the nitrogen anion, an excess electron is unbound, therefore no EA is presented for this case.
The conventional coupled-cluster contributions tend to be smaller with the increase of the excitation level. The exception here is the EA of fluorine, for which the contributions from full quadruples are larger than those obtained at the (Q) level of theory. Similar behavior can be observed in the case of the EA of oxygen where the (5) correction is larger that the full quadruples contribution. The F12 corrections are of the order of magnitude of a few meV for H, C and N. For the fluorine atom and for the EA of oxygen this correction is significantly larger and amounts to 35, 21 and 20 meV, respectively for the IP and EA of fluorine and the EA of oxygen. The scalar and one-electron Darwin relativistic corrections are of the order of magnitude of a few meV for the second-row atoms. The exceptions here are again fluorine and oxygen, where these contributions are significantly larger. The spin-orbit coupling corrections are of similar importance, exceptions here are large contributions to the IP of nitrogen and oxygen and the more significant contribution to the EA of fluorine. The two-electron Darwin term, as well as higher order relativistic contributions, such as the Breit interactions, are expected to be small for light atoms, therefore they have been neglected within the present work. The DBOC corrections are rather small in the case of the second-row atoms and, as expected, more significant for the hydrogen atom.
The final results compare well with the experimental values, except for the IP of oxygen and the EA of fluorine. For these cases we observe larger discrepancies. The IPs of hydrogen, carbon, nitrogen and oxygen agree to within a fraction of a meV with the experiment. The largest deviation here, associated with the IP of nitrogen, amounts to 0.0007 eV. The agreement with the experimental values for the EAs of hydrogen and carbon is similar. The most difficult cases among the considered atoms refer to oxygen and fluorine. The EA computed for oxygen differs from the experimental value by ca. 4 meV. The discrepancy associated with the EA of fluorine amounts to ca. 5 meV. This is most likely due to the incomplete convergence of the coupled-cluster hierarchy that is observed for these cases. The significant F12 contributions suggest that the basis set truncation errors are also larger for these cases, than for the other atoms.

7.4 Conclusions

The ionization potentials and electron affinities of the H, C, N, O and F atoms have been computed by means of state-of-the-art electronic structure methods. The conventional coupled-cluster calculations were performed up to the connected pentuple excitation level. For the purpose of the basis set truncation correction the implementation of the CCSD(F12) model in TURBOMOLE was applied. Final results were supplemented with relativistic and diagonal Born-Oppenheimer corrections. Estimated values of the IPs and EAs are in good agreement with the experimental values and the deviations do not exceed 0.7 meV, in the cases of H, C and N atoms and the IP of O atom. The results obtained for fluorine differ by ca. 1 and 5 meV from the experiment, respectively for the IP and EA. The EA of oxygen is plagued with discrepancy that amounts to ca. 4 meV.

Table 8: Exponents (in a_0^{-2}) of the diffuse Gaussians added to the aug-cc-pwCV5Z MP2-fitting basis sets[a].

ℓ	H	C	N	O	F
s	0.0440	0.0330	0.0410	0.0370	0.0510
	0.0180	0.0170	0.0200	0.0140	0.0210
	0.0071	0.0092	0.0096	0.0055	0.0086
p	0.0780	0.0360	0.0390	0.0690	0.0990
	0.0310	0.0150	0.0140	0.0280	0.0430
	0.0130	0.0059	0.0048	0.0120	0.0190
d	0.0930	0.0340	0.0430	0.0650	0.0880
	0.0370	0.0140	0.0170	0.0300	0.0430
	0.0150	0.0054	0.0065	0.0140	0.0210
f	0.0940	0.0550	0.0850	0.0920	0.1300
	0.0380	0.0220	0.0360	0.0350	0.0550
	0.0150	0.0090	0.0150	0.0140	0.0220
g	0.3200	0.0980	0.1400	0.1700	0.2500
	0.1300	0.0400	0.0590	0.0680	0.1100
	0.0510	0.0160	0.0240	0.0280	0.0460
h	0.4100	0.2600	0.3200	0.3600	0.4700
	0.1700	0.1500	0.1600	0.1700	0.2200
	0.0660	0.0800	0.0770	0.0760	0.1000
i		0.4100	0.4900	0.5900	0.7900
		0.2000	0.2200	0.2500	0.3400
		0.0980	0.0980	0.1100	0.1500

[a] aug-cc-pV5Z MP2-fitting basis for H.

Table 9: Exponents (in a_0^{-2}) of the diffuse Gaussians added to the aug-cc-pV5Z exchange-fitting basis sets.

ℓ	H	C	N	O	F
s	0.118	0.076	0.118	0.142	0.175
p	0.200	0.072	0.105	0.156	0.189
d	0.101	0.077	0.107	0.145	0.173
f	0.226	0.137	0.192	0.246	0.314
g	0.504	0.208	0.286	0.335	0.435
h	1.130	0.319	0.431	0.516	0.644
i		0.561	0.765	0.949	1.200

Table 10: Atomic ionization potentials (in eV).

	CCSD(T)[a]	(T)→(Q)[b]	(Q)→Q[c]	Q→(5)[d]	F12[e]	MVD[f]	SO[g]	DBOC[h]	Calc.	Exptl.[i]
H	13.60555	...	...	...	...	0.00018	...	−0.00741	13.59832	13.59843
C	11.25265	0.00567	0.00020	0.00000	0.00798	−0.00372	−0.00157	−0.00051	11.26070	11.26030
N	14.54185	0.00116	−0.00003	−0.00001	0.00957	−0.00617	−0.01102	−0.00057	14.53478	14.5341
O	13.57753	0.00501	−0.00006	−0.00003	0.03513	−0.00895	0.00967	−0.00047	13.61783	13.61805
F	17.40428	0.00185	−0.00017	0.00021	0.03526	−0.01277	−0.00414	−0.00050	17.42402	17.4228

[a] Calculated at the ae-CCSD(T)/d-aug-cc-pwCV5Z level.
[b] Calculated at the ae-CCSDT(Q)/d-aug-cc-pwCVQZ level.
[c] Calculated at the ae-CCSDTQ/d-aug-cc-pwCVTZ level.
[d] Calculated at the ae-CCSDTQ(5)/d-aug-cc-pwCVDZ level.
[e] Calculated at the ae-CCSD(F12)/d-aug-cc-pwCV5Z level.
[f] One-electron mass velocity and Darwin terms calculated at the ae-CCSD(T)/d-aug-cc-pwCV5Z level.
[g] Spin–orbit correction calculated from the experimentally observed spin–orbit splitting. See Ref. [159].
[h] Diagonal Born–Oppenheimer correction calculated at the ae-CCSD(T)/d-aug-cc-pwCV5Z level.
[i] Taken from Ref. [159].

Table 11: Atomic electron affinities (in eV).

	CCSD(T)[a]	(T)→(Q)[b]	(Q)→Q[c]	Q→(5)[d]	F12[e]	MVD[f]	SO[g]	DBOC[h]	Calc.	Exptl.
H	0.75167	...	...	...	0.00285	−0.00004	...	−0.00041	0.75407	0.75419[i]
C	1.25161	0.01308	−0.00027	−0.00030	0.00451	−0.00276	−0.00367	−0.00006	1.26214	1.26211[j]
O	1.43386	0.02013	−0.00373	−0.00483	0.02007	−0.00607	−0.00235	−0.00005	1.45703	1.46111[k]
F	3.40967	0.00106	−0.00249	−0.00706	0.02142	−0.00961	−0.01670	−0.00010	3.39619	3.4012[l]

[a] Calculated at the ae-CCSD(T)/d-aug-cc-pwCV5Z level.
[b] Calculated at the ae-CCSDT(Q)/d-aug-cc-pwCVQZ level.
[c] Calculated at the ae-CCSDTQ/d-aug-cc-pwCVTZ level.
[d] Calculated at the ae-CCSDTQ(5)/d-aug-cc-pwCVDZ level.
[e] Calculated at the ae-CCSD(F12)/d-aug-cc-pwCV5Z level.
[f] One-electron mass–velocity and Darwin terms calculated at the ae-CCSD(T)/d-aug-cc-pwCV5Z level.
[g] Spin–orbit correction calculated from the experimentally observed spin–orbit splitting. See Ref. [159]. Oxygen negative-ion spin–orbit data from Ref. [162].
[h] Diagonal Born–Oppenheimer correction calculated at the ae-CCSD(T)/d-aug-cc-pwCV5Z level.
[i] From Ref. [163]. See also Ref. [164].
[j] From Ref. [165]. See also Ref. [164].
[k] From Ref. [162]. See also Ref. [164].
[l] From Ref. [166]. See also Ref. [164].

8 Summary

In this thesis the implementation of the explicitly correlated coupled-cluster singles-and-doubles model (CCSD(F12)) in the TURBOMOLE package has been presented. This model is available within the **ricc2** [51] module that is an integral part of the TURBOMOLE [69] package. The program is capable of computing the CCSD(F12) energies with RHF, ROHF and UHF reference wave functions.

Among other implementations of the CCSD(F12) model, the one presented here is unique, because it combines integral-direct elements with density fitting techniques. The most demanding step of the CCSD model (the transformation involving the Coulomb integrals with four virtual indices) is performed using the AO-based integral-direct scheme, whereas other parts exploit density fitting techniques. The former feature allows for reducing the cost of the most time consuming operation within the CCSD scheme. With approximately the same cost of the transformation itself ($n_{\mathrm{vir}}^4 n_{\mathrm{occ}}^2$ vs. $N^4 n_{\mathrm{occ}}^2$ in the cases of fully MO- and partially AO-based algorithms, respectively) one can avoid the expensive AO-MO transformation of the Coulomb integrals to the virtual space. The $\Omega_{\mu i \nu j}^{BF}$ intermediate, which is a "key quantity" of the AO-based part of the conventional code, is extensively used also for the computation of explicitly correlated contributions. The major part of the residuals within Ansatz 1 is obtained by appropriate use of this intermediate together with explicitly correlated $V_{x,y}^{\mu,\nu}$ intermediate. These two quantities are the most important intermediates within the whole CCSD(F12) scheme. The remaining explicitly correlated contributions to the residuals, involving the F12-amplitudes $t_{i,j}^{p,q''}$, are obtained with density fitting techniques. They have been implemented by generalizing the conventional routines that are responsible for computing the C, D and E^1 contributions to the conventional residual. The structure of the conventional code allowed for the elegant introduction of CABS indices into the existing routines without major restructuring. They are treated as the natural extension of the conventional virtual space. This was possible because the storing and contraction schemes are organized such that the contraction dimension always refers to the occupied indices with the virtuals being the outer ones.

Recently, other approximated explicitly correlated coupled-cluster models have been proposed. They are obtained by neglecting some explicitly correlated contributions to the residuals and keeping the $c_{i,j}^{x,y}$ coefficients at the predetermined values, or by treating the whole F12 part in the perturbative manner. Within the present work, the complete CCSD(F12) model was implemented, as well as the diagonal orbital-invariant approximation proposed by Tew *et al.* and the modified approximation proposed by Adler *et al.*.

Each explicitly correlated contribution to the coupled-cluster expression was discussed, and the most important parts of the implementation were illustrated with schematic algorithms. Due to the fact that the code is capable of treating also UHF orbitals, the appropriate formulas in the spin-orbital formalism were also presented. The implementation of the CCSD(F12) model should be considered as the generalization of the existing conventional CCSD code. Therefore, some aspects of the conventional implementation, relevant for the present work, were also discussed.

The code was applied to problems that can be of the interest of general chemistry. The barrier heights of the reactions $CH_4 + CH_3^\bullet \rightleftarrows CH_3{\cdots}H{\cdots}CH_3^\bullet \rightarrow CH_3^\bullet + CH_4$ and $CH_4 + CH_3^\bullet \rightleftarrows H{\cdots}C_2H_6^\bullet \rightarrow H^\bullet + C_2H_6$ were determined using state-of-the-art quantum chemistry methods. These barrier heights and simple transition-state theory were used to compute the rate constants. The rate constants of the latter reaction were orders of magnitude

smaller than those used for the modeling of the pyrolysis of ethylene, acetylene and propylene. Based on this result, the reinvestigation of the role of this reaction in the proposed mechanism was suggested.
The implementation of the CCSD(F12) model was applied for the accurate determination of the atomization energies of the test set of molecules proposed by Bakowies. The estimated standard deviation from the ATcT experimental data amounts to ± 0.13 kJ/mol per valence electron in the molecule. The agreement with experiment was obtained without empirical parameters or extrapolations.
The electron affinities and ionization potential of the H, C, N, O and F atoms were computed by using conventional coupled-cluster methods supplemented with the explicitly correlated treatment at the CCSD(F12) level of theory. Agreement with experimental values of the order of magnitude of a fraction of meV was reached for hydrogen, carbon and nitrogen.
Within each application project, the CCSD(F12) energy calculation was only a part of the whole computational procedure. The accurate determination of the thermodynamic quantities requires the inclusion of various contributions. For instance, one should include the effects from the levels of theory that are far enough in terms of the hierarchy of the many-body models (CCSD, (T), CCSDT, (Q) and so on) but also other contributions are important (*e.g.* non-adiabatic and relativistic effects). Such composite approach was applied and, at each level of theory including the CCSD(F12) model, the largest possible basis set was used. This led to a very good agreement with experimental data.

9 Zusammenfassung (in German)

Im Rahmen dieser Arbeit wurde die Implementierung des explizit korrelierten CCSD(F12)-Modells in das Modul **Ricc2** [51] des Programmpakets TURBOMOLE [69] vorgestellt. Neben RHF-Rechnungen können, z.B. für Moleküle mit offenen Schalen, sowohl ROHF- als auch UHF-Berechnungen durchgeführt werden.

Im Vergleich zu anderen CCSD(F12)-Implementierungen hebt sich die hier vorgestellte dadurch hervor, dass sie integral-direkte Elemente mit *density fitting*-Techniken (DF) verbindet. Der teuerste Schritt einer konventionellen CCSD-Berechnung—die Berechnung und Verarbeitung von Coulomb-Integralen mit vier virtuellen Indizes—wird vollständig AO-basiert und integral-direkt durchgeführt. Hierdurch werden die Kosten für den teuersten Schritt innerhalb der CCSD-Rechnung deutlich gesenkt, da die teuren AO-MO Transformationen mit vier virtuellen Indizes vermieden werden. Nimmt man näherungsweise gleiche Kosten für die Transformationen an, ergeben sich Skalierungsverhalten von $n_{\text{vir}}^4 n_{\text{occ}}^2$ im Fall von MO- und $N^4 n_{\text{occ}}^2$ im Fall von teilweise AO-basierten Algorithmen. Für eine solche AO-basierte Implementierung des konventionellen CCSD-Modells stellen die $\Omega_{\mu i \nu j}^{BF}$-Intermediate zentrale Größen dar, so dass diese für die Implementierung des CCSD(F12)-Modells eine Schlüsselrolle einnehmen. So setzt sich der größte Teil der Residuen für Ansatz 1 hauptsächlich aus diesen Beiträgen und (explizit korrelierten) $V_{x,y}^{\mu,\nu}$-Intermediaten zusammen. Bei diesen beiden Größen handelt es sich also um die wichtigsten Intermediate innerhalb des gesamten CCSD(F12)-Programms. Die verbleibenden explizit korrelierten Beiträge (welche die F12-Amplituden enthalten) werden mit DF berechnet. Die Implementierung erfolgte in einer solchen Art, dass die vorhandenen Routinen, welche die Intermediate C, D und E^1 berechnen, erweitert wurden. Dies war möglich, da beim Speichern und Kontrahieren die virtuellen Orbitale stets als langsame Indizes verwendet werden. Somit konnten die CABS-Virtuellen elegant als natürliche Erweiterung des virtuellen Raumes ohne größere Umstrukturierungsarbeiten implementiert werden.

Im Rahmen der vorliegenden Arbeit wurde zunächst das gesamte CCSD(F12)-Modell implementiert. Kürzlich wurden genäherte Versionen des CCSD(F12)-Modells in der Literatur vorgeschlagen. Hierbei handelt es sich um Verfahren, bei denen Terme vernachlässigt werden, die F12-Koeffizienten unverändert bleiben, oder die gesamte F12-Korrektur störungstheoretisch erfolgt. Zwei dieser Vorschläge wurden ebenfalls im Rahmen der vorliegenden Arbeit implementiert, da—ausgehen vom CCSD(F12)-Modell—bestimmte Terme einfach übersprungen werden können. Bei den beiden genäherten Versionen handelt es sich um eine diagonale orbital-invariante Methode (vorgeschlagen von Tew *et al.*) sowie um das CCSD-F12b-Modell (vorgeschlagen von Adler *et al.*).

In der vorliegenden Arbeit wurde jeder explizit korrelierte Beitrag zu den CC-Gleichungen diskutiert. Darüberhinaus wurden die wichtigsten Teile der Implementierung anhand von schematischen Algorithmen illustriert. Aufgrund der Tatsache, dass der Code ebenfalls UHF-Referenzwellenfunktionen akzeptiert, wurden die Formeln im Spinorbital-Formalismus präsentiert. Wie weiter oben beschrieben, kann die CCSD(F12)-Implementierung aufgrund der Programmstruktur der bereits vorliegenden CCSD-Implementierung als Erweiterung dieses vorhandenen Codes verstanden werden (im Gegensatz zu einer denkbaren Umstrukturierung). Deshalb wurden manche Aspekte der konventionellen Implementierung, die für die aktuelle Arbeit relevant sind, ebenfalls diskutiert.

Nach der Implementierung erfolgte die Anwendung auf chemische Fragestellungen. Hierzu zählen Energiebarrieren für die Reaktionen $CH_4 + CH_3^\bullet \rightleftarrows CH_3{\cdots}H{\cdots}CH_3^\bullet \rightarrow CH_3^\bullet + CH_4$ und $CH_4 + CH_3^\bullet \rightleftarrows H{\cdots}C_2H_6^\bullet \rightarrow H^\bullet + C_2H_6$, welche mit modernen Methoden der Quan-

tenchemie berechnet wurden. Die so erhaltenen Energiebarrieren wurden mit der Theorie des Übergangszustandes kombiniert, um Reaktionsgeschwindigkeitskonstanten zu berechnen. Diese waren für die zweite Reaktion um Größenordnungen kleiner als solche, die in der Literatur verwendet wurden, um die Pyrolyse von Ethylen, Acetylen und Propylen zu modellieren. Aufgrund dieser Ergebnisse wurde vorgeschlagen, die Rolle dieser Reaktion in den Modellierungen zu überprüfen.
Die zweite Anwendung der CCSD(F12)-Implementierung stellt die Berechnung hochgenauer Atomisierungsenergien dar. Dies wurde am Testsatz von 105 Molekülen durchgeführt, welcher von Bakowies vorgeschlagen wurde. Hierbei zeigte sich eine Standardabweichung von ±0.13 kJ/mol pro Valenzelektron von den experimentellen ATcT-Daten. Diese herrausragende Übereinstimmung mit dem Experiment wurde ohne empirische Parameter oder Extrapolationen erhalten.
Weiterhin wurden Elektronenaffinitäten und Ionisierungspotentiale der Atome Wasserstoff, Kohlenstoff, Stickstoff, Sauerstoff und Fluor mit konventionellen Coupled-Cluster-Methoden berechnet, unterstützt durch CCSD(F12)-Ergebnisse. Eine sehr gute Übereinstimmung innerhalb von einem meV mit experimentellen Daten wurde für Wasserstoff, Kohlenstoff und Stickstoff erhalten.
Innerhalb jeder Anwendung stellt die CCSD(F12)-Berechnung nur eine von mehreren Rechnungen dar. Um thermodynamische Größen hochgenau berechnen zu können, müssen Beiträge unterschiedlicher Art berücksichtigt werden. So muss zum einen die Wellenfunktion einen ausreichenden Mehrteilchencharakter haben (CCSD, (T), CCSDT, (Q), usw.), zum anderen müssen relativistische oder nicht-adiabatische Korrekturen behandelt werden. In der vorliegenden Arbeit wurden diese verschiedenen Beiträge nicht nur berücksichtigt, sondern darüberhinaus wurde stets die größtmögliche Basis verwendet. Durch diese Vorgehensweise ist es möglich, hochgenaue Ergebnisse zu erhalten, die mit experimentellen Daten verglichen werden können.

10 Podsumowanie (in Polish)

W niniejszej pracy doktorskiej przedstawiona została implementacja jawnie skorelowanej metody sprzężonych klasterów (CCSD(F12)) w programie TURBOMOLE. Implementacja tego modelu jest dostępna w module **ricc2** [51], który z kolei jest integralną częścią pakietu TURBOMOLE [69]. Przy użyciu tego programu obliczyć można energie CCSD(F12) z referencyjnymi funkcjami falowymi RHF, UHF i ROHF.

Zaprezentowana implementacja jest unikalna wśród innych realizacji modelu CCSD(F12), ponieważ łączy ona elementy "integral-direct" z technikami fitowania gęstości. Najbardziej złożony krok modelu CCSD (transformacja angażująca całki kulombowskie z czterema indeksami wirtualnymi) jest wykonywany przy użyciu schematu "integral-direct" częściowo w bazie atomowej. Pozostałe części wykorzystują techniki fitowania gęstości. Ta pierwsza cecha pozwala na zmniejszenie kosztu najbardziej czasochłonnej operacji w całym schemacie CCSD. Przy utrzymaniu porównywalnego kosztu samej transformacji ($n_{\mathrm{vir}}^4 n_{\mathrm{occ}}^2$ vs. $N^4 n_{\mathrm{occ}}^2$ odpowiednio dla algorytmów zrealizowanych w bazach orbitali molekularnych i atomowych) można uniknąć kosztochłonnej transformacji do bazy orbitali wirtualnych. Wielkość pośrednia $\Omega^{BF}_{\mu i \nu j}$, która jest kluczowym elementem implementacji konwencjonalnego modelu CCSD w bazie atomowej, jest też szeroko wykorzystywana do obliczania jawnie skorelowanych wkładów. Dla Ansatzu 1 znacząca część członów w równaniach amplitudowych jest uzyskiwana przez właściwe użycie wielkości pośredniej $V^{\mu,\nu}_{x,y}$. Te dwie wielkości są kluczowe dla całego schematu CCSD(F12). Pozostałe jawnie skorelowane wkłady do równań amplitudowych uzyskane są przy zastosowaniu technik fitowania gęstości. Zostały one zaimplementowane przez uogólnienie istniejących procedur odpowiedzialnych za obliczanie członów C, D i E^1 do konwencjonalnych równań amplitudowych. Struktura kodu konwencjonalnego modelu CCSD pozwoliła na bardzo eleganckie wprowadzenie indeksów CABS do istniejących procedur bez znacznych modyfikacji programu. Indeksy te traktowane są jako naturalne rozszerzenie bazy wirtualnej. Było to możliwe dzięki specyficznemu zorganizowniu schematów kontrakcji. Wymiary kontrakcji związane są tutaj z orbitalami obsadzonymi, a indeksy wirtualne nie są jawnie zaangażowane w kontrakcje.

Ostanio zaproponowane zostały inne przybliżenia jawnie skorelowanych modeli sprzężonych klasterów. Uzyskane one zostały przez zaniedbanie pewnych jawnie skorelowanych wkładów do równań amplitudowych i ustalenie wartości współczynników $c^{x,y}_{i,j}$. Inne propozycje sugerują traktowanie całej jawnie skorelowanej części w sposób perturbacyjny. W niniejszej pracy zaimplementowany został kompletny model CCSD(F12) jak również dwa inne, przybliżone modele.

Każdy jawnie skorelowany wkład został przedyskutowany, a najważniejsze elementy implementacji zilustrowane zostały schematycznymi algorytmami. W związku z tym, że kod obsługuje także orbitale UHF, odpowiednie wyrażenia zaprezentowane zostały także w formaliźmie spinoorbitalnym. Implementacja modelu CCSD(F12) powinna być tutaj rozumiana jako uogólnienie istniejącego kodu CCSD. Dlatego też pewne aspekty implementacji metody konwencjonalnego CCSD, istotne z punktu widzenia tej pracy, zostały także przedyskutowane.

Program zastosowany został do problemów, które dotyczą szeroko rozumianej chemii ogólnej. Za pomocą nowoczesnych metod chemii kwatnowej wyznaczone zostały wysokości barier reakcji chemicznych $CH_4 + CH_3^{\bullet} \rightleftarrows CH_3 \cdots H \cdots CH_3^{\bullet} \rightarrow CH_3^{\bullet} + CH_4$ oraz $CH_4 + CH_3^{\bullet} \rightleftarrows H \cdots C_2H_6^{\bullet} \rightarrow H^{\bullet} + C_2H_6$. Otrzymane wartości barier, w połączeniu z teorią stanu przejściowego, zostały wykorzystane do obliczenia stałych szybkości reakcji. Stałe szybkości drugiej reakcji były kilka rzędów wielkości mniejsze niż te użyte do modelowania pirolizy etylenu, acetylenu i propylenu. Bazując na tych wynikach zasugerowane zostało powtórne przenalizowanie roli tej reakcji w zaproponowanych mechanizmach.

Implementacja modelu CCSD(F12) zastosowana została do precyzyjnego wyznaczenia energii atomizacji cząsteczek ze zbioru testowego zaproponowanego przez Bakowiesa. Oszacowane odchylenie standardowe od wartości eksperymentalnych ATcT wyniosło ±0.13 kJ/mol na każdy elektron walencyjny w cząsteczce. Zgodność z eksperymentem uzyskana została bez konieczności stosowania parametrów empirycznych bądź technik ekstrapolacyjnych.

Przy pomocy technik sprzężonych klasterów i jawnie skorelowanej metody CCSD(F12) policzone zostały powinowactwa elektronowe i potencjały jonizacji atomów H, C, N, O i F. Zgodność z wartosciami doświadczalnymi rzędu ułamka meV została osiągnięta dla wodoru, węgla i azotu.

W każdym projekcie aplikacyjnym jawnie skorelowane obliczenia energii CCSD(F12) były tylko częścią całej procedury obliczeniowej. Dokładne wyznaczenie wielkości termodynamicznych wymaga uwzględnienia wielu czynników. Powinny zostać wzięte pod uwagę, na przykład, wkłady pochodzące z poziomów teorii, ktore są wystrczająco zaawansowane w hierarchii modeli wielociałowych (CCSD, (T), CCSDT, (Q) i tak dalej), ale również inne efekty są istotne (np. efekty nieadiabatyczne i relatywistyczne). Taki "kompozytowy" model został zaproponowany w ramach niniejszej pracy. Na każdym rozważanym poziomie teorii, włączając w to CCSD(F12), największa z możliwych baz funkcyjnych została użyta. Doprowadziło to do uzyskania bardzo dobrej zgodności z eksperymentem.

A Acronyms

ABS	Auxiliary basis set
ATcT	Active thermochemical tables
An1	Ansatz 1
An3	Ansatz 3
AO	Atomic orbital
BCH	Baker-Campbell-Hausdorff
CABS	Complementary auxiliary basis set
CC	Coupled-cluster
CCSD	Coupled-cluster singles-and-doubles
CCSD-F12	Full explicitly correlated coupled-cluster singles-and-doubles
CCSD(F12)	Linearized explicitly correlated coupled-cluster singles-and-doubles
CI	Configuration interaction
CVD	Chemical vapor deposition
DBOC	Diagonal Born-Oppenheimer correction
DF	Density fitting
DIIS	Direct inversion in the iterative subspace
EA	Electron affinity
ECG	Exponentially correlated Gaussian
FCI	Full configuration interaction
HF	Hartree-Fock
IP	Ionization potential
LCG	Linear combination of Gaussian
MO	Molecular orbital
MP2	Second order Møller-Plesset perturbation theory
MP2-F12	Explicitly correlated second order Møller-Plesset perturbation theory
RI	Resolution of the identity
RHF	Restricted Hartree-Fock
ROHF	Restricted open-shell Hartree-Fock

STG Slater-type geminal

UHF Unrestricted Hartree-Fock

ZPVE Zero-point vibrational energy

B List of scientific publications

1. M. Haranczyk, R.A. Bachorz, J. Rak, M. Gutowski, D. Radisic, S.T. Stokes, J.M. Nilles, and K.H. Bowen,
Excess electron attachment induces barrier-free proton transfer in binary complexes of uracil with H_2Se and H_2S but not with H_2O,
J. Phys. Chem. **B 107**, 7889 (2003).

2. R.A. Bachorz and J. Komasa,
Variational calculations on H_4^{2+} using exponentially correlated Gaussian wave functions,
Comp. Meth. Sci. Tech. **11**, 5 (2005).

3. R.A. Bachorz, J. Rak, and M. Gutowski,
Stabilization of very rare tautomers of uracil by an excess electron,
Phys. Chem. Chem. Phys. **7**, 2116 (2005).

4. R.A. Bachorz, M. Haranczyk, I. Dabkowska, J. Rak, and M. Gutowski,
Anion of the formic acid dimer as a model for intermolecular proton transfer induced by a π^* excess electron,
J. Chem. Phys. **122**, 204304 (2005).

5. J.E. Jaffe, R.A. Bachorz, and M. Gutowski,
Low-temperature polymorphs of ZrO_2 and HfO_2: A density-functional theory study,
Phys. Rev. **B 72**, 144107 (2005).

6. M. Gutowski, R.A. Bachorz, T. Autrey, and J.C. Linehan,
Relative stability of $(NH_3BH_3)2$, $[NH_3\text{-}BH_2\text{-}NH_3]^+BH_4^-$, and $[BH_3\text{-}NH_2\text{-}BH_3]^-NH_4^+$,
Prep. Pap. - Am. Chem. Soc., Div. Fuel Chem. **50 (2)**, 496 (2005).

7. K. Mazurkiewicz, R.A. Bachorz, M. Gutowski, and J. Rak,
On the unusual stability of valence anions of thymine based on very rare tautomers: A computational study,
J. Phys. Chem. **B 110**, 24696 (2006).

8. R.A. Bachorz, W. Klopper, and Maciej Gutowski,
Coupled-cluster and explicitly correlated perturbation-theory calculations of the uracil anion,
J. Chem. Phys. **126**, 085101 (2007).

9. J.E. Jaffe, R.A. Bachorz, and M. Gutowski,
Band offset and magnetic property engineering for epitaxial interfaces: a monolayer of M_2O_3 (M = Al, Ga, Sc, Ti, Ni) at the α-Fe2O3/α-Cr2O3 (0001),
Phys. Rev. **B 75**, 205323 (2007).

10. R. Leist, J. A. Frey, P. Ottiger, H.-M. Frey, S. Leutwyler, R.A. Bachorz, and W. Klopper,
Nucleobase-fluorobenzene interactions: Hydrogen bonding wins over π-stacking,
Angew. Chem. Int. Ed. **46**, 7449 (2007).

11. X. Li, K.H. Bowen, M. Haranczyk, R.A. Bachorz, K. Mazurkiewicz, J. Rak, and M. Gutowski,
Photoelectron spectroscopy of adiabatically bound valence anions of rare tautomers of the nucleic acid bases,
J. Chem. Phys. **127**, 174309 (2007).

12. J. Rak, K. Mazurkiewicz, M. Kobylecka, P. Storoniak, M. Haranczyk, I. Dabkowska, R.A. Bachorz, M. Gutowski, D. Radisic, S.T. Stokes, S.N. Eustis, D. Wang, X. Li, Y.J. Ko, and K.H. Bowen,
Radiation induced molecular phenomena in nucleic acids: A comprehensive theoretical and experimental Analysis,
edited by M.K. Shukla and J. Leszczynski (Springer, Netherlands).

13. S.N. Eustis, D. Radisic, K.H. Bowen, R.A. Bachorz, M. Haranczyk, G.K. Schenter, and M. Gutowski
Electron-driven acid-base chemistry: proton transfer from hydrogen chloride to ammonia,
Science **319**, 936 (2008).

14. R.A. Bachorz, F.A. Bishoff, S. Höfener, W. Klopper, P. Ottiger, R. Leist, J.A. Frey, and S. Leutwyler
Scope and limitations of the SCS-MP2 method for stacking interactions,
Phys. Chem. Chem. Phys. **10**, 2758 (2008).

15. R.A. Bachorz, M. Gutowski, W. Klopper, X. Li, and K.H. Bowen,
Photoelectron spectrum of valence anions of uracil and first-principles calculations of excess electron binding energies,
J. Chem. Phys. **129**, 054309 (2008).

16. P. Ottiger, C. Pfaffen, R. Leist, S. Leutwyler, R.A. Bachorz, and W. Klopper,
Strong N-H...π hydrogen bonding in amide-benzene interactions,
J. Phys. Chem. **B**, **113**, 2937 (2009).

17. R.A. Bachorz, M. Haranczyk, G. Lupica, and M. Gutowski,
Electrostatic potential maps of damaged DNA studied by image analysis tools. 8-Oxoguanine and abasic site lesions,
J. Mol. Model. **15**, 815 (2009).

18. W. Klopper, R.A. Bachorz, D.P. Tew, J. Aguiliera-Iparraguirre, Y. Carissan, and C. Hättig,
Accurate coupled-cluster calculations of the reaction barrier heights of two $CH_3^{\bullet}+CH_4$ reactions,
J. Phys. Chem. **A**, in press.

19. R.A. Bachorz, W. Cencek, R. Jaquet, and J. Komasa,
Rovibrational energy levels of H_3^+ with energies above the barrier to linearity,
J. Chem. Phys. **131**, 024105 (2008).

C Acknowledgements

It was a great pleasure for me to spend couple of years at the Karlsruhe University. I would like to express my gratitude to my supervisor Prof. Wim Klopper for his guidance, constant encouragement, constructive criticism and support throughout the course of this work.

I would sincerely like to thank Prof. Christof Hättig from Ruhr-Universität Bochum for his involvement in this work. His contribution and help are very important to me.

I am thankful to Dr. David Tew. The hours of "debug sessions" spent with him are very important part of this work. Thank you Dave!

I would also like to thank each and everyone of my lab members for their helpful discussions and for creating a very nice environment in the lab.

Finally I would like to thank my mother. She always believes in my abilities and she is always with me in difficult moments of my live. Dziękuję mamo!

A special word to my girlfriend Ania. Thank you for being with me, thank you for your smile.

References

[1] G. Herzberg and A. Monfils, *J. Mol. Spectrosc.* **5**, 482 (1960).

[2] E. Schrödinger, *Ann. Phys.* **79**, 361 (1926).

[3] T. Kato, *Commun. Pure Appl. Math.* **10**, 151 (1957).

[4] R. T. Pack and W. B. Brown, *J. Chem. Phys.* **45**, 556 (1966).

[5] T. Helgaker, P. Jørgensen, and J. Olsen, *Molecular Structure Theory*, John Wiley & Sons, Chichester, 2000.

[6] E. A. Hylleraas, *Z. Phys.* **48**, 469 (1929).

[7] E. A. Hylleraas, *Z. Phys.* **54**, 347 (1929).

[8] H. M. James and A. S. Coolidge, *J. Chem. Phys.* **1**, 825 (1933).

[9] W. Kołos and C. C. Roothaan, *Rev. Mod. Phys.* **32**, 219 (1960).

[10] W. Kołos and L. Wolniewicz, *J. Chem. Phys.* **41**, 3663 (1964).

[11] W. Kołos and L. Wolniewicz, *J. Chem. Phys.* **43**, 2429 (1965).

[12] G. Herzberg, *J. Mol. Spectrosc.* **33**, 147 (1970).

[13] S. F. Boys, *Proc. Roy. Soc.* **A 258**, 402 (1960).

[14] K. Singer, *Proc. Roy. Soc.* **A 258**, 412 (1960).

[15] J. Rychlewski (Ed.), *Explicitly Correlated Wave Functions in Chemistry and Physics Theory and Applications*, Kluwer Academic Publishers, 2003.

[16] W. Klopper and W. Kutzelnigg, *Chem. Phys. Lett.* **134**, 17 (1987).

[17] O. Sinanoglu, *J. Chem. Phys.* **36**, 706 (1962).

[18] O. Sinanoglu, *J. Chem. Phys.* **36**, 3198 (1962).

[19] P. Wind, W. Klopper, and T. Helgaker, *Theor. Chem. Acc.* **107**, 173 (2002).

[20] E. F. Valeev, *Chem. Phys. Lett.* **395**, 190 (2004).

[21] F. R. Manby, *J. Chem. Phys.* **119**, 4607 (2003).

[22] C. Villani and W. Klopper, *J. Phys.* **B 38**, 2555 (2005).

[23] T. B. Adler, H.-J. Werner, and F. R. Manby, *J. Chem. Phys.* **130**, 054106 (2009).

[24] S. Ten-no, *J. Chem. Phys.* **121**, 117 (2004).

[25] S. Ten-no, *J. Chem. Phys.* **126**, 014108 (2007).

[26] J. Noga and P. Valiron, *Chem. Phys. Lett.* **324**, 166 (2000).

[27] E. F. Valeev, *J. Chem. Phys.* **125**, 244106 (2006).

[28] R. A. Bachorz, W. Klopper, and M. Gutowski, *J. Chem. Phys.* **126**, 085101 (2007).

[29] S. Ten-no, *Chem. Phys. Lett.* **447**, 175 (2007).

[30] G. Knizia and H.-J. Werner, *J. Chem. Phys.* **128**, 154103 (2008).

[31] D. P. Tew, W. Klopper, C. Neiss, and C. Hättig, *Phys. Chem. Chem. Phys.* **9**, 1921 (2007).

[32] D. P. Tew, W. Klopper, C. Neiss, and C. Hättig, *Phys. Chem. Chem. Phys.* **10**, 6325 (2008).

[33] D. R. Hartree, *Proc. Cam. Phil. Soc.* **24**, 89 (1927).

[34] W. Fock, *Z. Phys.* **61**, 126 (1930).

[35] J. Cižek, *J. Chem. Phys.* **45**, 4256 (1966).

[36] T. D. Crawford and H. F. Schaefer III, An introduction to coupled cluster theory for computational chemists, in *Reviews in Computational Chemistry*, edited by K. B. Lipkowitz and T. R. Cundari, volume 14, page 33, VCH Publishers, New York, 2000.

[37] G. D. Purvis III and R. J. Bartlett, *J. Chem. Phys.* **76**, 1910 (1982).

[38] J. Noga and R. J. Bartlett, *J. Chem. Phys.* **86**, 7041 (1987).

[39] G. E. Scuseria and H. F. Schaefer III, *Chem. Phys. Lett.* **152**, 382 (1988).

[40] J. Noga and W. Kutzelnigg, *J. Chem. Phys.* **101**, 7738 (1994).

[41] H. Fliegl, W. Klopper, and C. Hättig, *J. Chem. Phys.* **122**, 084107 (2005).

[42] W. J. Lauderdale, J. F. Stanton, J. Gauss, J. D. Watts, and R. J. Bartlett, *J. Chem. Phys.* **97**, 6606 (1992).

[43] P. J. Knowles, J. S. Andrews, R. D. Amos, N. C. Handy, and J. A. Pople, *Chem. Phys. Lett.* **186**, 130 (1991).

[44] D. P. Tew and W. Klopper, *J. Chem. Phys.* **123**, 074101 (2005).

[45] S. Ten-no, *Chem. Phys. Lett.* **398**, 56 (2004).

[46] W. Klopper and J. Almlöf, *J. Chem. Phys.* **99**, 5167 (1993).

[47] W. Klopper and C. C. M. Samson, *J. Chem. Phys.* **116**, 6397 (2002).

[48] C. Hättig, unpublished, 2009.

[49] F. A. Bischoff, S. Höfener, A. Glöß, and W. Klopper, *Theor. Chem. Acc.* **121**, 11 (2008).

[50] S. Höfener, F. A. Bischoff, A. Glöß, and W. Klopper, *Phys. Chem. Chem. Phys.* **10**, 3390 (2008).

[51] C. Hättig and F. Weigend, *J. Chem. Phys.* **113**, 5154 (2000).

[52] P. Pulay, *Chem. Phys. Lett.* **73**, 393 (1980).

[53] A. Szabo and N. S. Ostlund, *Modern Quantum Chemistry*, Dover Publications, INC., Mineola, New York, 1996.

[54] BLAS routines, http://www.netlib.org/blas.

[55] C. Møller and M. S. Plesset, *Phys. Rev.* **A 46**, 618 (1934).

[56] H. Koch, A. S. de Merás, T. Helgaker, and O. Christiansen, *J. Chem. Phys.* **104**, 4157 (1996).

[57] C. Neiss and C. Hättig, *J. Chem. Phys.* **126**, 154101 (2007).

[58] R. Ahlrichs, M. Bär, M. Häser, H. Horn, and C. Kölmel, *Chem. Phys. Lett.* **162**, 165 (1989).

[59] T. Shiozaki, M. Kamiya, S. Hirata, and E. F. Valeev, *J. Chem. Phys.* **129**, 071101 (2008).

[60] D. P. Tew, W. Klopper, and C. Hättig, *Chem. Phys. Lett.* **452**, 326 (2008).

[61] T. B. Adler, G. Knizia, and H.-J. Werner, *J. Chem. Phys.* **127**, 221106 (2007).

[62] G. Knizia, T. B. Adler, and H.-J. Werner, *J. Chem. Phys.* **130**, 054104 (2009).

[63] K. Norinaga and O. Deutschmann, Ind. Eng. Chem. Res. **46**, 3547 (2007).

[64] K. Tabayashi and S. H. Bauer, Combust. Flame **34**, 63 (1979).

[65] L. S. Kassel, J. Am. Chem. Soc. **57**, 833 (1935).

[66] G. B. Skinner and R. A. Ruehrwein, J. Phys. Chem. **64**, 1736 (1959).

[67] R. A. Back, Can. J. Chem. **61**, 916 (1983).

[68] R. A. Bachorz, C. Hättig, W. Klopper, and D. P. Tew, unpublished, 2009.

[69] Turbomole, local experimental version of v5.10, Turbomole GmbH, Karlsruhe. For the current public version, see http://www.turbomole.com.

[70] J. F. Stanton, J. Gauss, J. D. Watts, and R. J. Bartlett, *J. Chem. Phys.* **94**, 4334 (1991).

[71] J. Noga, S. Kedžuch, J. Šimunek, and S. Ten-no, *J. Chem. Phys.* **128**, 174103 (2008).

[72] A. Köhn, G. W. Richings, and D. P. Tew, *J. Chem. Phys.* **129**, 201103 (2008).

[73] E. F. Valeev, *Phys. Chem. Chem. Phys.* **10**, 106 (2008).

[74] D. Bokhan, S. Ten-no, , and J. Noga, *Phys. Chem. Chem. Phys.* **10**, 3320 (2008).

[75] M. Torheyden and E. F. Valeev, *Phys. Chem. Chem. Phys.* **10**, 3410 (2008).

[76] D. Bokhan, S. Bernadotte, and S. Ten-no, *Chem. Phys. Lett.* **469**, 214 (2009).

[77] J. Noga, S. Kedžuch, J. Šimunek, and S. Ten-no, *J. Chem. Phys.* **130**, 029901 (2009).

[78] T. Shiozaki, M. Kamiya, S. Hirata, and E. F. Valeev, *J. Chem. Phys.* **130**, 054101 (2009).

[79] A. Köhn, *J. Chem. Phys.* **130**, 131101 (2009).

[80] K. Raghavachari, G. W. Trucks, J. A. Pople, and M. Head-Gordon, *Chem. Phys. Lett.* **157**, 479 (1989).

[81] J. D. Watts, J. Gauss, and R. J. Bartlett, *Chem. Phys. Lett.* **200**, 1 (1992).

[82] For H, the aug-cc-pwCVTZ basis is identical with the aug-cc-pVTZ basis.

[83] J. F. Stanton *et al.*, ACES II, (Mainz–Austin–Budapest 2005 version), a quantum chemical program package for high-level calculations of energies and properties, containing the integral packages MOLECULE (J. Almlöf and P. R. Taylor), PROPS (P. R. Taylor) and ABACUS (T. Helgaker, H. J. Aa. Jensen, P. Jorgensen, and J. Olsen). For the current version, see http://www.cfour.de (CFOUR program).

[84] J. P. Layfield, M. D. Owens, and D. Troya, *J. Chem. Phys.* **128**, 194302 (2008).

[85] S. M. Remmert, S. T. Banks, and D. C. Clary, *J. Phys. Chem. A* **113**, 4255 (2009).

[86] J. D. Watts and R. J. Bartlett, *J. Chem. Phys.* **93**, 6104 (1990).

[87] Y. J. Bomble, J. F. Stanton, M. Kállay, and J. Gauss, *J. Chem. Phys.* **123**, 054101 (2005).

[88] M. Kállay and J. Gauss, *J. Chem. Phys.* **123**, 214105 (2005).

[89] MRCC, a string-based quantum chemical program suite written by M. Kállay. See http://www.mrcc.hu.

[90] M. Kállay and P. R. Surján, *J. Chem. Phys.* **115**, 2945 (2001).

[91] K. A. Peterson, T. B. Adler, and H.-J. Werner, *J. Chem. Phys.* **128**, 084102 (2008).

[92] K. E. Yousaf and K. A. Peterson, *J. Chem. Phys.* **129**, 184108 (2008).

[93] C. Hättig, *Phys. Chem. Chem. Phys.* **7**, 59 (2005).

[94] F. Weigend, *J. Comp. Chem.* **29**, 167 (2007).

[95] T. H. Dunning, Jr., *J. Chem. Phys.* **90**, 1007 (1989).

[96] R. A. Kendall, T. H. Dunning, Jr., and R. J. Harrison, *J. Chem. Phys.* **96**, 6796 (1992).

[97] K. A. Peterson and T. H. Dunning, Jr., *J. Chem. Phys.* **117**, 10548 (2002).

[98] D. Feller, *J. Comp. Chem.* **17**, 1571 (1996).

[99] K. L. Schuchardt *et al.*, *J. Chem. Inf. Model.* **47**, 1045 (2007).

[100] http://tyr0.chem.wsu.edu/~kipeters/basissets/basis.html.

[101] G. Tarczay, A. G. Császár, W. Klopper, and H. M. Quiney, *Mol. Phys.* **99**, 1769 (2001).

[102] S. Coriani, T. Helgaker, P. Jørgensen, and W. Klopper, *J. Chem. Phys.* **121**, 6591 (2004).

[103] J. E. Sansonetti and W. C. Martin, *J. Phys. Chem. Ref. Data* **34**, 1559 (2005).

[104] D. A. McQuarrie and J. D. Simon, *Physical Chemistry: A Molecular Approach*, University Science Books, Sausalito, 1997.

[105] J. I. Steinfeld, J. S. Francisco, and W. L. Hase, *Chemical Kinetics and Dynamics*, University Science Books, Prentice Hall, New York, 2nd edition, 1999.

[106] M. L. Strekalov, *Chem. Phys.* **355**, 62 (2009).

[107] J. Aguilera-Iparraguirre, A. D. Boese, W. Klopper, and B. Ruscic, *Chem. Phys.* **346**, 56 (2008).

[108] W. Klopper, B. Ruscic, D. P. Tew, F. A. Bischoff, and S. Wolfsegger, *Chem. Phys.* **356**, 14 (2009).

[109] B. Ruscic *et al.*, *J. Phys. Chem. A* **108**, 9979.

[110] B. Ruscic, *Active Thermochemical Tables*, 2005 Yearbook of Science and Technology (annual update to McGraw-Hill Encyclopedia of Science and Technology), University Science Books, McGraw-Hill, New York, 2004.

[111] B. Ruscic *et al.*, *J. Phys. Conf. Ser.* **16**, 561 (2005).

[112] B. Ruscic *et al.*, *J. Phys. Chem.* **110**, 6592 (2006).

[113] D. W. Schwenke, *Spectrochim. Acta* **A 55**, 731 (1999).

[114] D. W. Schwenke, *Spectrochim. Acta* **A 58**, 849 (2002).

[115] A. Karton, B. Ruscic, and J. M. L. Martin, *J. Mol. Struct. THEOCHEM* **811**, 345 (2007).

[116] A. Karton, E. Rabinovich, J. M. L. Martin, and B. Ruscic, *J. Chem. Phys.* **125**, 144108 (2006).

[117] T. Helgaker, W. Klopper, H. Koch, and J. Noga, *J. Chem. Phys.* **106**, 9639 (1997).

[118] M. E. Harding *et al.*, *J. Chem. Phys.* **128**, 114111 (2008).

[119] N. Kungwan and T. N. Truong, *J. Phys. Chem. A* **109**, 7742 (2005).

[120] J. A. Kerr and M. J. Parsonage, *Evaluated Kinetic Data on Gas Phase Hydrogen Transfer Reactions of Methyl Radicals*, Butterworths, London, 1976.

[121] D. G. Truhlar, A. D. Isaacson, R. T. Skodje, and B. C. Garrett, *J. Phys. Chem.* **86**, 2252 (1982).

[122] L. A. Curtis, K. Raghavachari, G. W. Trucks, and J. A. Pople, *J. Chem. Phys.* **94**, 7221 (1991).

[123] L. A. Curtis, K. Raghavachari, and P. C. Redfern, *J. Chem. Phys.* **109**, 7764 (1998).

[124] L. A. Curtis, P. C. Redfern, and K. Raghavachari, *J. Chem. Phys.* **126**, 084108 (2007).

[125] N. J. DeYonker, T. R. Cundari, and A. K. Wilson, *J. Chem. Phys.* **124**, 114104 (2006).

[126] N. J. DeYonker *et al.*, *J. Chem. Phys.* **125**, 104111 (2006).

[127] J. J. A. Montgomery, M. J. Frisch, J. W. Ochterski, and G. A. Petersson, *J. Chem. Phys.* **110**, 2822 (1999).

[128] J. J. A. Montgomery, M. J. Frisch, J. W. Ochterski, and G. A. Petersson, *J. Chem. Phys.* **112**, 6532 (2000).

[129] A. L. L. East and W. D. Allen, *J. Chem. Phys.* **99**, 4638 (1993).

[130] A. G. Császár, W. D. Allen, and H. F. Schaefer III, *J. Chem. Phys.* **108**, 9751 (1998).

[131] P. L. Fast, M. L. Sánchez, and D. G. Truhlar, *Chem. Phys. Lett.* **306**, 407 (1999).

[132] P. L. Fast, J. Corchado, M. L. Sánchez, and D. G. Truhlar, *J. Phys. Chem. A* **103**, 3139 (1999).

[133] P. L. Fast, J. Corchado, M. L. Sánchez, and D. G. Truhlar, *J. Phys. Chem. A* **103**, 5129 (1999).

[134] A. Tajti *et al.*, *J. Chem. Phys.* **121**, 11599 (2004).

[135] Y. J. Bomble *et al.*, *J. Chem. Phys.* **125**, 064108 (2006).

[136] J. M. L. Martin and G. de Oliveira, *J. Chem. Phys.* **111**, 1843 (1999).

[137] A. D. Boese *et al.*, *J. Chem. Phys.* **120**, 4129 (2004).

[138] A. Karton, P. R. Taylor, and J. M. L. Martin, *J. Chem. Phys.* **127**, 064104 (2007).

[139] D. Bakowies, *J. Chem. Phys.* **127**, 084105 (2007).

[140] D. E. Woon and T. H. Dunning, Jr., *J. Chem. Phys.* **103**, 4572 (1995).

[141] F. Weigend and R. Ahlrichs, *Phys. Chem. Chem. Phys.* **7**, 3279 (2005).

[142] B. Ruscic, unpublished results obtained from Active Thermochemical Tables (ATcT) ver. 1.36 using the Core (Argonne) Thermochemical Network ver. 1.070, 2008.

[143] D. E. Woon and T. H. Dunning, Jr., *J. Chem. Phys.* **98**, 1358 (1993).

[144] J. F. Stanton, C. L. Lopreore, and J. Gauss, *J. Chem. Phys.* **108**, 7190 (1998).

[145] J. D. Watts, J. Gauss, and R. J. Bartlett, *J. Chem. Phys.* **98**, 8718 (1993).

[146] C. L. Jansen and I. M. B. Nielsen, *Chem. Phys. Lett.* **290**, 423 (1998).

[147] W. Kutzelnigg, *Theor. Chim. Acta* **68**, 445 (1985).

[148] W. Klopper and W. Kutzelnigg, *Chem. Phys. Lett.* **134**, 17 (1987).

[149] W. Kutzelnigg and W. Klopper, *J. Chem. Phys.* **94**, 1985 (1991).

[150] W. Klopper, R. Röhse, and W. Kutzelnigg, *Chem. Phys. Lett.* **178**, 455 (1991).

[151] J. Noga, W. Kutzelnigg, and W. Klopper, *Chem. Phys. Lett.* **199**, 497 (1992).

[152] W. Klopper and J. Noga, *J. Chem. Phys.* **103**, 6127 (1995).

[153] J. Noga, P. Valiron, and W. Klopper, *J. Chem. Phys.* **115**, 2022 (2001).

[154] A. J. May, E. Valeev, R. Polly, and F. R. Manby, *Phys. Chem. Chem. Phys.* **7**, 2710 (2005).

[155] H.-J. Werner and F. R. Manby, *J. Chem. Phys.* **124**, 054114 (2006).

[156] F. R. Manby, H.-J. Werner, T. B. Adler, and A. J. May, *J. Chem. Phys.* **124**, 094103 (2006).

[157] D. Yamaki, H. Koch, and S. Ten-no, *J. Chem. Phys.* **127**, 144104 (2007).

[158] H.-J. Werner, T. B. Adler, and F. R. Manby, *J. Chem. Phys.* **126**, 164102 (2007).

[159] See http://physics.nist.gov/PhysRefData/.

[160] See http://physics.nist.gov/cuu/constants/.

[161] DALTON, a molecular electronic-structure program, release 2.0 (2005), see http://www.kjemi.uio.no/software/dalton/dalton.html.

[162] D. M. Neumark, K. R. Lykke, T. Andersen, and W. C. Lineberger, *Phys. Rev.* **A 32**, 1890 (1985).

[163] K. R. Lykke, K. K. Murray, and W. C. Lineberger, *Phys. Rev.* **A 43**, 6104 (1991).

[164] Nist chemistry webbook, gas phase ion energetics data, available at http://webbook.nist.gov.

[165] M. Scheer, R. C. Bilodeau, C. A. Brodie, and H. K. Haugen, *Phys. Rev.* **A 58**, 2844 (1998).

[166] C. Blondel, C. Delsart, and F. Goldfarb, *Phys. Rev.* **A 32**, L281 (2001).

Made in United States
Cleveland, OH
05 November 2026

47659895R00063